Cattle
Beet
Capital

Cattle
Beet
Capital

**Making
Industrial
Agriculture
in Northern
Colorado**

MICHAEL WEEKS

University of Nebraska Press
LINCOLN

Portions of chapters 4 and 5 first appeared as "Measuring
Expertise: Ralph Parshall and Watershed Management,
1910–1940," in *The Greater Plains: Rethinking a Region's
Environmental Histories*, ed. Brian Frehner and
Kathleen A. Brosnan (University of Nebraska Press,
2021), 179–201, and "Sugar State: Industry, Science,
and Nation in Colorado's Sugar Beet Fields," *Western
Historical Quarterly* 48 (October 2017): 367–91.

The University of Nebraska Press is part of a land-
grant institution with campuses and programs on the
past, present, and future homelands of the Pawnee,
Ponca, Otoe-Missouria, Omaha, Dakota, Lakota, Kaw,
Cheyenne, and Arapaho Peoples, as well as those of the
relocated Ho-Chunk, Sac and Fox, and Iowa Peoples.

Library of Congress Cataloging-in-Publication Data
Names: Weeks, Michael (Historian), author.
Title: Cattle beet capital: making industrial agriculture
in Northern Colorado / Michael Weeks
Other titles: Making industrial agriculture
in Northern Colorado
Description: Lincoln: University of Nebraska Press,
[2022] | Includes bibliographical references and index.
Identifiers: LCCN 2022000527
ISBN 9781496208415 (hardback)
ISBN 9781496232304 (epub)
ISBN 9781496232311 (pdf)
Subjects: LCSH: Cattle—Feeding and feeds—Colorado. |
Cattle trade—Colorado. | Beef industry—Colorado. |
BISAC: HISTORY / United States / State & Local / West
(AK, CA, CO, HI, ID, MT, NV, UT, WY) | BUSINESS
& ECONOMICS / Industries / Agribusiness
Classification: LCC HD9433.C6 W44 2022 |
DDC 338.1/76209788—dc23/eng/20220208
LC record available at https://lccn.loc.gov/2022000527

Set in Minion Pro by Laura Buis.
Designed by L. Auten.

For William (Bill) Unrau: scholar, mentor, friend

CONTENTS

ILLUSTRATIONS

ACKNOWLEDGMENTS

This project was founded in curiosity but made possible by a supportive community of scholars, friends, and professionals. My debts to them are only superseded by my gratitude for their generosity. What follows is a modest attempt to recognize their contributions.

When I began this project in 2013, I was fortunate enough to have resources available locally while completing a doctorate at the University of Colorado. Archivist David Hays offered guidance and perspective as I stumbled through material. Since much of my research delves into Colorado's agricultural history, I soon migrated to the archives at Colorado State. There, Patricia Rettig and Linda Meyers patiently asked about my research, often suggesting sources I would not have found otherwise. Among the many library assistants at CSU, I want to give a hearty thanks to Clarissa Trapp. The early help of these skilled professionals enabled ideas to germinate that might have remained buried.

During early and late stages of the project, local libraries and museums helped immensely. I found rich materials on sugar beet history at city museums in Loveland and Fort Morgan and on cattle feeding in Greeley. Peggy Ford provided guidance as I looked into the history of Monfort of Colorado and directed me to interviewees. In the project's final stages, when Covid-19 limited travel, Katalyn Lutkin digitized several dozen pictures for my use. Several of these appear in the book. I also benefited from materials at History Colorado and Denver Public Library where Wendell Cox contributed enthusiasm and knowledge.

As the project expanded, I was aided by grants and fellowships that broadened my thinking and research. A Roaring Fork Fellowship and Sears Grant from the University of Colorado enabled travel to Washington DC, where I found a treasure trove of materials on USDA scientists. Generous grants from the Hagley Museum and the National Science Foundation funded invaluable time in Delaware and Philadelphia focused on agricultural chemicals. A generous grant from the Poudre Heritage Alliance supported critical research on the irrigation plumbing that slakes Colorado's thirsty fields.

In 2019 I received a Samuel Hays Fellowship from the American Society for Environmental History that enabled me to interview many people in Northern Colorado for this book. Their stories inform large sections of the final two chapters. They invited me into their homes, met me for coffee, and took me on tours. I cannot do their aid justice, but I can share their names: John Stencil, Kent Peppler, Thomas Trout, Bill Hamerich, Chuck Sylvester, Kathleen Sullivan-Kelley, Richard Maxfield, John Matsushima, Greg Ludlow, Gilbert Barela, Jerry Sonnenberg, Jim Miller, Don Ament, Greg Ludlow, Del Miles, William R. Farr, Rod Ulrich, Richard Seaworth, Mike Callicrate, Robert Zimdahl, Ray Chamberlain, and Gordon Kruse.

I was fortunate to have a fantastic corps of advisors and peers during my graduate career. Michael Amundson at Northern Arizona first encouraged me to pursue a PhD in 2007. Phoebe Young, Thomas Andrews, and Mark Fiege gave thoughtful feedback and advice while this project was a dissertation. Most importantly, Paul Sutter gave copious feedback for multiple drafts and was a fount of insight into the field of history. Years later, he still replies to emails faster than anyone I know. Among the fantastic group of peers who provided ready feedback and needed distractions, Karen Lloyd D'Onofrio, Sara Porterfield, and Doug Sheflin deserve special mention.

The camaraderie of my peers at Utah Valley University and beyond have aided this project in its final stages. To that end, I would like to thank Mark Lentz, Keith Snedegar, Hilary Hungerford, and Brendan McCarthy. I have specifically benefited from the feedback of Scott Abbott, Lyn Bennett, Adrienne Winans, Jenna Nigro, John Hunt, and Michael Goode who read parts of this manuscript and offered timely thoughts. I am also grateful for the help of Bart Elmore who provided feedback on the final stages of the book and friendly banter along the way.

Increasingly I have learned the value of professional editors. Parts of chapters 4 and 5 previously appeared in the *Western Historical Quarterly* and as a chapter in *The Greater Plains: Rethinking a Region's Environmental Histories.* To that end, I am grateful for the work of David Rich Lewis, Anne Hyde, Kathleen Brosnan, and Brian Frehner. Finally, I cannot say enough about the skilled and tireless work of Bridget Barry at the University of Nebraska Press. There is a direct correlation between her patience with this project and my sanity at its conclusion.

I am grateful for two individuals who blurred lines between the academic and the personal. Philip Riggs has been a sounding board for this project from the start, capable of seamlessly transitioning between big academic ideas and our personal lives. He is also the author of several maps and visuals in the book. Bill Unrau also deserves more credit than I can give. Bill is a retired historian who I met when starting graduate school. He befriended me, offering advice and needed perspective through graduate school and beyond. Bill passed away just months after I completed doctoral studies. He still inspires.

In 2004 I rekindled a love for history while teaching an Advanced Placement U.S. History class to high school students in Placerville, California. Observing this, my wife, Sacha, suggested that I consider graduate school. Since then, there have been more physical moves and life transitions than I care to enumerate. This project has seen the light of day because of her encouragement then, and in the years since. If there is any grace in this book, I have her to thank.

Cattle
Beet
Capital

Introduction

Few companies in the American West articulate the agricultural transformations of the twentieth century more completely than Monfort
of Colorado. During the early 1970s, its Greeley-based operations
possessed the greatest concentration of commercially fed cattle in
the world. Monfort fattened, slaughtered, boxed, and shipped over
half a million animals each year. The company and its peers dominated regional agriculture, contracting with farmers to cultivate more
than three hundred thousand tons of corn grain and silage annually,
thereby enabling penned steers and heifers to pack on nearly three
pounds of heft daily. These impressive gains depended on a complex
irrigation infrastructure that diverted every drop of water within the
South Platte River watershed and tunneled the headwaters of the
Colorado River beneath the Continental Divide to slake industry
thirsts. Monfort's appetites extended beyond its Northern Colorado
headquarters in purchasing synthetic chemicals, fertilizers, pharmaceuticals, and machinery from far-flung research labs and factories.
The company also contracted with multinational chemical and fossil
fuel corporations to synthesize the plastics and dry ice used to pack,
box, cool, and ship its beef across the globe. Greeley residents could
not ignore Monfort since it was as recognizable by its smell as by its
beef. Hundreds of thousands of cattle packed into feedlots produced
copious wastes whose odors invaded olfactory senses up and down
the Front Range of the Rocky Mountains. Some simply referred to
the odor as the smell of Greeley while, for others, it was the scent of
profit. Either way, Greeley smelled like industrial agriculture.[1]

Before I stepped foot in Colorado, Monfort was both foreign and
familiar to me. I grew up in a rural town south of the San Fran-

cisco Bay Area where my family raised goats, ducks, a horse, and more litters of Cocker Spaniel puppies than I can recall. During my adolescent years, we fattened and slaughtered several pigs and a steer on our modest acre of land. Regular chores included feeding animals, cleaning their pens, and scooping scattered piles of manure for deposit in a compost pile that functioned to replenish the nitrogen and phosphorous content in a garden that supplied our family of five with most of its vegetables. I experienced animals and their manure as both repelling and regenerative. By contrast, commercial feedlots were an oddity that revolted, and yet peaked, my curiosity. During frequent road trips to Southern California along the I-5 Freeway, I marked the midpoint by the smell of Harris Ranch—several roadside miles of beef cattle packed into commercial feedlots whose concentrated excrement saturated the air in our modest vehicle. Though we temporarily held our noses, for me, the idea of Harris Ranch always lingered long after its scent faded. The vast chasm between backyard animals and roadside cattle left much unexplained.

I lived within whiffing distance of Monfort's Greeley feedlots from 1997 to 1999, while teaching high school in Commerce City, Colorado. As the regional home to food processing and agricultural corporations, as well as the region's most prominent oil refinery, Commerce City played the role of industrial stepchild to nearby Denver, which was rapidly developing a reputation as the hub of outdoorsy hipsters who were fast erasing the city's cow-town past. Greeley's commercial feedlot odors permeated Commerce City on days when prevailing winds blew from the north or when inversions stagnated airflow. Despite its pervasiveness, the smell of Greeley was not foreign to Commerce City, since it mingled with other industrial outputs—sulfur from the oil refinery and meat byproducts being transformed into dog chow at the local Purina factory.

Eleven years later, when I arrived in Boulder, Colorado, to study environmental history, Greeley's aroma took on a different air. When its feedlot smells descended on Boulder, an upscale haven of liberal politics and environmental activism, my pungent curiosity evolved

Fig. 1. Aerial photograph of Monfort commercial cattle feedlots near Greeley, ca. 1970. The facility held a hundred thousand steers at full capacity. City of Greeley Museums. Used by permission.

into academic study. I wanted to understand how this place, which defined industrial meat production and was a metaphor for olfactory putrescence, had come to be. In the process, I discovered a compelling story that traces the arc of modern food production in America.

It was not until a pleasant spring day in 2014 that I finally viewed Monfort's feedlots up close. Though the company has changed hands several times in the last half century, its Greeley area feedlots remain in much the same form as in the early 1970s. Driving northward on one of the many back roads that traverse Weld County, I was impressed at how the landscape mirrored stereotypes of rural America. Stalks of corn, cut low to the ground, remained in several fields, awaiting removal in advance of spring planting. Pastures with horses and livestock mingled with fields of alfalfa. Full irrigation canals flowed with purpose. The sights and smells of rural America, mixed with the warm sun and pleasant breezes, drew me in.

Approaching Greeley from the south, the pastoral yielded to the industrial. At first, I noticed a fetid smell that grew in strength as I drove. This was not the smell of manure in pastures, but of concentrated excrement. Shortly, the stench took material form. In countless pens, each occupying nearly an acre, a cohort of identical steers crowded in front of feed bunks where they chewed, snorted, and coughed their way through a meal. They stood on ground that was uniformly brown and lifeless, offering a stark contrast to the vast fields and pastures, which, until a few moments ago, had dominated my senses. In the center of the pens sat two landmarks. One was a water trough whose liquid was pumped from underground wells. The other was an amorphous mound of manure that grew and receded as workers cleaned pens. In one dystopic scene, a steer stood atop one of these fecal hills as if to proclaim temporary lordship over his pen mates. Traveling beyond the feedlots, a tall white building came into focus. Initially, it appeared to be an industrial-sized grain silo. I later discovered that this was a feed mill calibrated by engineers, nutritionists, and technicians to mix the grains, roughage, vitamins, minerals, and antibiotics that supplied daily rations for one hundred thousand animals. The machines, feeds, pens, and cattle I observed that day—in Monfort's feedlots and countless others like it—engineered the bulk of America's beef.

At first glance, this scene bore no resemblance to the cooperative agricultural venture that settled on the site of Monfort's feedlots in 1870. Named the Union Colony by its founders, *New York Tribune* editors Horace Greeley and Nathan Meeker, it pooled resources from several hundred would-be pioneers for a planned settlement. The Union Colony's blueprint for success required that all members contribute labor and financial resources to build an irrigation system that would enable market crops to flourish on small acreage. Unlike Greeley in 1970, cattle played little part. Though some settlers possessed livestock, colony records indicate that cattle were few and that they largely grazed on the open range surrounding the colony.

Moreover, concerns over livestock often had more to do with how to fence them out of farms than confine them within. Settlers struggled to cultivate viable crops, fought with neighboring towns, and eked out a modest living. A year after settlers founded the Union Colony, they named their town after Horace Greeley, who contributed money, encouragement, and editorial space during its infancy. The town's name and its location seem to be all that the Union Colony shares with modern Greeley.[2]

What transformed the modest Union Colony of 1870 into the agro-industrial juggernaut of Greeley in 1970? This book is animated by that question. In the early stages of the project, I sketched a narrative arc that presented as a fall from grace. The villains of the story were big agricultural corporations that employed capital and resources to exploit a population of small farmers incapable of resisting. They harnessed the region's land, labor, and water to their purposes, destroying ecological niches that evolved over millennia. It was a clear story of declension whereby unrestrained abuse of people and the environment by a small cohort of capitalists resulted in a denuded landscape and a disempowered population.

I confess that this is the story I wanted to find. During my spare time, I struggle to grow my own food. I frequent farmers' markets and pay close attention to food origins and methods of cultivation. The works of Michael Pollan, Barbara Kingsolver, and Wendell Berry occupy prime bookshelf space in my home.[3] Words like *synthetic, processed, artificial,* and *high fructose* are anathema to me. So, when I encountered Greeley's feedlots, I was immediately repelled, but not so much by the smell. They somehow violated the intimacy I desire with the food I eat and the land on which it is produced, replacing them with remote industrial processes designed to dominate nature rather than cooperate with it. I confess to being a romantic when it comes to farming. Being deeply suspicious of industrial agriculture, it was easy to envision a research project that mapped onto my own sensibilities.

The narrative edifice I constructed crumbled under the weight of the economic, environmental, and relational forces I encountered.

Though wealth accumulation through land and water speculation were among the factors that motivated Anglo settlement, I discovered that, well into the twentieth century, power had not concentrated noticeably. Small, middle-class landowners were the norm, and irrigation water was widely distributed among them. Even when Great Western Sugar, the nation's largest sugar beet refiner, brought vast capital investments into Northern Colorado, the company could not dominate the region's resources. Corporate executives, growers, and state-sponsored scientists learned that sugar beets could not be cultivated on the same land year in and year out since mining the region's soil eroded its health and decimated harvests. Instead, they found that working with existing crop rotations and mixed husbandry supported both sustained profits and a healthy agroecology. Where labor was concerned, nature pushed back against capitalism. Even as regional growers exploited families of migrant laborers to perform the tasks associated with cultivation during the first half of the twentieth century, their attempts to mechanize operations were foiled by biological limitations beyond their control. It became clear that concentrated capital could not so easily bend agriculture to its will and that organisms—human or otherwise—often blunted market forces seeking to rationalize them.[4]

But where were the cattle and the feedlots? Whenever they appeared, they existed in the background during the first half of the twentieth century—consuming sugar beet byproducts, providing fertilizer in the form of manure, and offering meager income for farmers who fattened modest numbers of livestock during the winter. How then did this cattle-fattening side business become Northern Colorado's industrial juggernaut? It turned out that answers were found as much in the lab as in the field. Developments in plant breeding, chemistry, pharmaceuticals, and veterinary medicine made it possible to fatten cattle year-round on inexpensive feeds in tight spaces. This is where I came to one of my most important realizations. The story of modern agriculture in Northern Colorado was as much about developments outside the region as those within it. Farmers, laborers, and corporate agriculture were bound up together with laboratories and manufacturers of agricultural implements and chemicals. Relationships were not linear, but networked, and the story about a particular place was also about many places.

Greeley is the hub of a region that historical geographer William Wyckoff calls the Northern Colorado Piedmont. A subregion of the Great Plains that extends eastward from the Front Range of the Rocky Mountains, its topography and historical trajectory are most closely allied with the Arkansas River Valley, its twin to the south.[5] The physical geography of the Piedmont includes the South Platte River, its Front Range tributaries, and the irrigated lands adjacent to these waters. The South Platte emerges from the Rocky Mountains southwest of Denver, winds its way through the city, and then arcs in a northeasterly direction through the state until it merges with the North Platte River in Nebraska. As it meanders away from Denver, the South Platte collects the waters of several other streams, including Boulder Creek, St. Vrain Creek, and Big Thompson and Cache la Poudre Rivers. On average, the Piedmont receives less than fourteen inches of precipitation annually and the watershed on which it depends delivers a paltry 1.6 million acre-feet of water annually, making it one of the nation's drier watersheds.[6] It is a region of contrasts—defined by aridity and energized by water. The Piedmont is also distinguished by easy access to transportation. Anglo settlers who arrived on the Piedmont after the Civil War lobbied for and situated their towns adjacent to rail lines that could take their goods locally to Denver and each of the most populated Colorado towns that lined the eastern front of the Rocky Mountains. Piedmont towns were also linked to the Union Pacific Railroad, which freighted their goods anywhere in the United States. These transportation arteries and the water infrastructure that Anglo settlers built transformed the Piedmont into one of the American West's most rapidly settled agricultural landscapes. By 1900, few regions in the West were as well suited to market agriculture as the Northern Colorado Piedmont.[7]

Aridity and irrigated agriculture defined the post–Civil War development of the American West, and the Piedmont is an essential region for understanding both.[8] Most states in the West apportion their

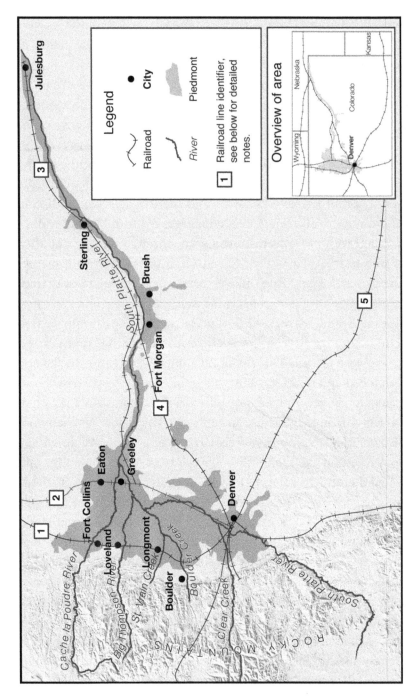

Fig. 2. Map of the Northern Colorado Piedmont. Created by Philip Riggs.

water according to the Doctrine of Prior Appropriation, a series of statutes and legal precedents that allotted water to individual users based on the date of settlement and the amount of water they could put to "beneficial use." Those laws crystallized as Piedmont settlers clashed over the right to divert water from local streams. Yet, while much has been written about prior appropriation, little has been said about the agricultural development of the region that pioneered Western water law after 1880.[9] The Piedmont remained among the West's most important centers of irrigation into the twentieth century. Until 1920, Colorado vied with California for the most irrigated acreage in the nation, and the Piedmont possessed far and away the greatest quantity of irrigated land in the state.[10] Water is an essential actor throughout this story, informing settlement patterns, land use, politics, and crop choice. As historians such as Richard White have highlighted, watersheds do work. The South Platte River watershed was overburdened with labor. Thus, my work brings clarity to discussions of water in the American West.[11]

Cattle Beet Capital offers a counterpoint to studies of agricultural development in the West, which are dominated by California.[12] Unlike most of the arid West, California water law was a hybrid of prior appropriation and riparian rights, and many of its water projects, such as the Central Valley Project, were engineered by the Federal Bureau of Reclamation. Piedmont water infrastructure, like water developments in most of the West, was far more local and decentralized, with Reclamation playing only a minor role for much of its history. In addition, while heavily capitalized speculators quickly snapped up many of California's most productive agricultural lands prior to the Civil War, their histories are not analogous to much of the West, where colder climates, smaller populations, and moderate investment resulted in a more prolonged development of industrial agriculture.[13] Between 1900 and 1960, California grew from the tenth leading producer of agricultural products in the United States to number one—a position it has held ever since. No other western state has cracked the top ten.[14] While this offers an argument for California's importance, it suggests that the state's agricultural development is not representative of the rest of the American West. In that sense, examining agriculture on the Piedmont—where environmental limits, crop selections, and smaller capital investments narrowed options and slowed development—provides a better model for understanding western agriculture.[15]

Though *Cattle Beet Capital* is rooted in a specific region in the American West, its geographic scope and methodology extend beyond it. As a work of environmental and agricultural history, it uses the transformation of the Piedmont from 1870 to 1970 as a template for analyzing how humans rationalized and commodified specific landscapes within the context of capitalism and its institutions. During the twentieth century, the Piedmont was at the leading edge of the production of two global commodities: sugar and beef. For most of the period between 1900 and 1950, the Piedmont sacked more sucrose than any other U.S. region. Farmers oriented their crops around sugar beet cultivation, and one corporation, Great Western Sugar, monopolized the processing of every last beet. The cheap ubiquity of sugar in American life depended on armies of poorly paid migrant laborers whose work and home lives were inseparable from the fields in which they toiled. Thus, *Cattle Beet Capital* offers a thorough analysis of the evolving relationships between capitalism, labor, and the agroecology humans constructed to support them.

During the post–World War II era, the nation's per capita consumption of beef doubled and commercial cattle feeding supplanted sugar beets as the Piedmont's principal economic driver.[16] Regional feeders were the first to combine breakthroughs in the chemical and biological sciences with the mechanization of animal feeding in confined spaces. These processes reengineered cattle lifecycles, altered human relationships with livestock, and reconfigured the region's agroecology. The Piedmont economy revolved heavily around beef and sugar during the twentieth century. I place their commodification within the dynamic set of environmental, human, technological, and scientific relationships occurring at the landscape level.[17]

Despite an emphasis on commodities, *Cattle Beet Capital* is not a history of consumption. This is intentional and contains an implied argument throughout. While consumer markets for sugar, beef, and other commodities informed Piedmont crop choices, land use was determined by a host of other factors such as available water, soil health, agronomic knowledge, and the influence of state and industrial actors. Answers to questions about cultivation prior to World War II were more often found in soils than in retail establishments. Even after the war, when agro-industrial goods eroded traditional practices, farmers and feeders concerned themselves with the chemicals, pharmaceuticals, fertilizers, and feeds purchased for consumption on the farm before considering the faceless consumers on the other end of the supply chain. Landscapes such as the Piedmont provide a location-specific barometer for understanding how modern food systems evolved.

Historians of capitalism and industrial agriculture have questioned the degree to which nature has been simply a vehicle for capital accumulation and whether capitalism was historically compatible with sustaining healthy farm ecologies.[18] Evidence supplied by Piedmont agriculture suggests that answering that question depends on the crops being grown, how and where capital and labor were apportioned, and the period being analyzed. I argue that prior to World War II the agroecology developed by Piedmont farmers was largely sustainable due to its dependence on crop and livestock commodities that reinforced soil health and supported economic stability while limiting speculative wealth. Even when the beet sugar industry reoriented the region's economy, Great Western Sugar and its growers were restrained by the idiosyncrasies of the sugar beet, a cultivar that was most lucrative when farmers cycled the byproducts of their own farms back into the soil. In this I concur with historian Colin Duncan who argues that for capitalism to promote a healthy landscape, producers must depend on locally available resources, "giving them a built-in incentive to watch over that part of the environment in which they are actively interfering."[19] The period from 1930 to 1970, on the other hand, illustrates a slow erosion and then total collapse of Piedmont agroecology as farmers and cattle feeders embraced mechanical, biological, and petrochemical technologies.[20] These changes were energized by rapid industrial developments and by state-sponsored researchers who replaced fundamental scientific questions with field trials for agro-industry. The Greeley feedlots of 1970 represent both a fully articulated industrial agriculture and the culmination of robust efforts to rationalize and replace natural processes occurring in soils, crops, and animal bodies.

The market agriculture practiced by Piedmont farmers from 1870 to 1930—the subject of chapters 1 and 2—achieved limited success in large part through experimentation, irrigation, and by embracing advances in relevant agricultural sciences. While farmers marketed their crops outside the region and imported some livestock and seeds, their cultivation practices were limited by Piedmont water and soils. In chapter 2, I examine the ascendancy of corporate agriculture after 1900 in the form of the beet sugar industry, demonstrating that its success was largely a function of bolstering existing agricultural practices to accommodate the new cultivar. Economic gain operated in concert with sustaining the region's agroecology, since the corporation and its growers maintained a dependent relationship on Piedmont resources.

Labor presents both a critical exception to reliance on local resources and the most important catalyst for change. Chapter 3 shows how the lived experiences of sugar beet workers were baked into Piedmont agroecology. Sugar beet cultivation was embedded in the soils, bodies, homes, and education of its workers. In this, I agree with historian Linda Nash who argues that the agricultural work environment is often inseparable from the bodies of its workers.[21] The grueling and repetitive nature of sugar beet labor placed entire families side by side in the fields and generated recognizable impacts on their bodies. Their homes, situated adjacent to fields, were constructed haphazardly within the habitat of organisms that thrived in the region's irrigated fields. The imprints of child labor were immediately recognizable in their education since they arrived late from the harvest and left early to thin beets. Injuries, poor nutrition, and difficulty concentrating supported preexisting stereotypes that enabled the system to essentialize Mexicans as lazy and intellectually inferior. While contingent factors brought a marginalized class of laborers into the fields, their toil was inseparable from the Piedmont's constructed agroecology.

Scientists within various agencies of the U.S. Department of Agriculture (USDA) and the complex of agricultural colleges played a significant and underappreciated role in the development of Piedmont agriculture. They are the subjects of chapters 4 and 5. Where they shared common goals and agendas and, broadly speaking, they usually did—I employ the term *agri-state* to refer to this complex of scientists and officials. Agri-state employees encompassed the only publicly funded conglomeration of scientists and officials whose mission was to serve the needs of farmers specifically and American agriculture more generally. I argue that they largely failed in their mission. As scientists from agro-industry promised greater efficiency and expanded yields through science and technology, agri-state officials failed to question potential harms. Rather, they claimed that the products of industry, when used as directed, were unqualified goods. This philosophy led agri-state researchers and scientists to emphasize industry field trials over substantive research. In the process, they became junior partners with agro-industry and the site of agronomic research expertise moved from public science on the Piedmont to private industry in distant labs. Further, by placing unquestioned faith in efficiency and expanded yields through technology, state-sponsored scientists supported a knowledge economy whereby performing experiments aimed at uncovering the long-term impacts of applying chemicals to the land or questioning the value of industrial products in agriculture had become largely taboo. I argue that by offering unflinching support for industry agri-state scientists and officials failed to safeguard the farmers and the lands that were essential to their founding mission.[22]

Cattle Beet Capital correlates agricultural practices with environmental change over time in a particular region of the American West. While this may seem elemental, it brings together periods that historians often separate. It is common to end studies of western agriculture during the 1920s and 1930s, or to focus primarily on the period following World War II.[23] This yields a fragmented picture since many of the technological and scientific innovations that transformed agricultural practice were developed between 1900 and 1945 but were adopted en masse following the war. Thus, it is easy to view the earlier period as characterized by limits on environmental impacts and the latter through the lens of environmental degradation.[24] *Cattle Beet Capital* shows how established agroecology held back market and technological forces prior to World War II, even as scientific, industry, and state interests developed the halting infrastructure to overwhelm it. Evidence of that developing infrastructure is foregrounded in chapters 2 through 5.

Efforts to decouple Piedmont agriculture from dependence on locally available resources were fully manifested in the decades following World War II and are explored in chapters 6 and 7. The combined efforts of agri-state, industry, and leaders of burgeoning commercial cattle feeding operations pushed farmers and feeders to divorce their operations from local and on-the-farm resources. Advances in plant breeding; an explosion of synthetic chemicals, antibiotics, and hormones; and new technologies for managing and feeding cattle in tightly packed feedlots ruptured former relationships and turned farmers and feedlot operators into consumers of industrial products emerging from distant factories and labs. To sustain yields, they demanded capital to solve problems once controlled by cycling nutrients back into the soil and rotating crops. Modern agriculture demands heavy investments. The Piedmont example shows that dependence on local landscapes sustained healthy agroecologies while divorcing commodity production from that dependence, yielded exploitation.

In many ways, the journey from the Union Colony in 1870 to Monfort in 1970 has enabled me to connect the metaphorical dots separating the backyard of my youth from the feedlots of Harris Ranch. My backyard seemed legible because I could observe the growth and decay that made food possible. In truth, however, this was a site filled with billions of complex interactions between organisms, minerals, and water. My perception of simplicity belied nature's complexities. By contrast, the feedlots at Harris Ranch and Greeley seemed foreign because machinery, chemicals, and biological engineering obscured simplicity. What I did not understand at the time was that the baffling infrastructure that engineered modern beef was not designed to harness nature's complexities so much as to eliminate them as factors in food production. Thus, the path from the Union Colony to Monfort traces the inversely proportional relationship between simplicity and complexity in nature that defines how a society feeds itself. It is a historical response to the question of where our food comes from that is at once unique to the Piedmont and applicable to agricultural landscapes everywhere.

1 Cultivating a Regional Agroecology

The South Platte was the most miserable river in the West, a trickle in the summer when its water was needed, a raging torrent in the spring. It was muddy, often more island than river, and prior to the introduction of irrigation, it had never served a single useful purpose in its halting career.

—JAMES MICHENER

On December 23, 1869, an expectant New York crowd gathered to hear the *New York Tribune*'s agricultural editor, Nathan Meeker, describe a pioneering opportunity. Meeker sought adventurous and hardy souls to join him in forming an irrigated agricultural settlement in the Colorado Territory the following year. Meeker was encouraged in his ambitions by his friend and *Tribune* editor, Horace Greeley, who agreed to provide financial support and print space to support the endeavor. While Greeley is best known for his reformist politics and writings during the 1850s and 1860s, and a failed run for the presidency in 1872, he was also a farming enthusiast who wrote extensively of his scientific experiments in cultivation on his hobby farm in upstate New York. Like Meeker, he believed that planned agricultural settlements in the West could thrive economically while modeling democratic values to the nation. So, when Meeker proposed the idea of a Colorado colony, Greeley responded that Meeker should "take hold of it." Owing to shared support for the North during and following the Civil War, the two agreed to name it the Union Colony. Approximately five hundred people showed up to Meeker's initial meeting in 1869. In less than two years there were over one thousand settlers.[1]

The agriculture that Union colonists fashioned from 1870 to 1900 shaped the Piedmont landscape for the foreseeable future. Settlers built intensive cultivation into the fabric of their founding ideals, pouring their resources into farming small plots of land aimed at the sustained cultivation of high value crops.[2] Colony founders allotted a portion of their initial investment in the endeavor to build irrigation works, intended to water modest acreage, believing that the land would yield abundant harvests if only sufficient water were available.[3] The original 1870 colony bylaws stated that colonists could select five-acre plots of land adjacent to town, or larger plots of up to forty acres located further afield. Sections of up to 160 acres were available for purchase from the Denver Pacific Railroad or out of the public domain through the Homestead Act. As farms quickly radiated outward from watercourses, Piedmont residents developed complex engineering schemes and legal frameworks to irrigate additional lands while storing water in reservoirs for diversion during the drier months of the year.[4] Intensive cultivation was also made possible through rails. Union colonists favored their settlement site over others due to its proximity to the Denver Pacific Railroad, since the growing city of Denver already provided a healthy market for regional crops. As the Piedmont grew, farmers and local boosters successfully lobbied for a spur rail line that would parallel the South Platte River, connecting to the Union Pacific in Wyoming. By 1900 rail lines crisscrossed the Piedmont. Emphasis on small acreage, irrigation, and access to markets attracted more settlers and, by 1900, the Northern Colorado Piedmont's productive, well-watered, and densely organized system of farming had laid down deep roots.[5]

Despite the communitarian goals of Nathan Meeker and Horace Greeley, Piedmont settlers emphasized an ethos of individualism and wealth accumulation with surprising consequences. Irrigation increased land values and supported speculation as some original colonists pocketed profits by selling their lands to newcomers, thus increasing the number of settlers with no attachment to founding colony goals.

Land speculation and water development fostered an urgency that attracted outside capital to tap Piedmont streams. In the meantime, open-range ranching attracted speculative ranchers who freighted millions of cattle into Northern Colorado, hoping to get rich quick by fattening cattle on the abundant grasses in the public domain.[6] Though unsustainable in the long run, they helped to reshape the region's agro-ecology as Piedmont crop farmers incorporated modest cattle herds (and other livestock) into their farming operations and remaining cattle ranchers learned to manage their herds through reliance on cultivated crops. Ironically, the collapse of one of the most speculative industries in American history contributed to a more sustainable agriculture by diversifying the region's farming and economic base.

Practical and theoretical agricultural science also explains evolving Piedmont farming prior to 1900. Evidence from the writings of settlers in the region show that they read and incorporated the research of seed and livestock breeders, chemists, and biologists. These were incorporated into evolving crop rotations and into the eventual adoption of crops such as clover, alfalfa, and other legumes that fixed nitrogen into the soil and served as a nutritious cattle-fattening additive. These developments were augmented by federal funding to develop experiment stations at agricultural colleges throughout the nation. Colorado's first experiment station was in the Piedmont town of Fort Collins on the campus of Colorado Agricultural College (CAC). Beginning in 1889, the college performed experiments that mirrored the needs of Piedmont farmers. Its researchers published farmers' bulletins, conducted demonstrations, and held classes for farmers and ranchers to help them apply the latest science to their livelihoods. While it is possible to overstate their pre-1900 influence, there is no question that CAC scientists became influential members of the Piedmont farming community, supporting the intensive agriculture that characterized the region.[7]

The Piedmont before 1870

While the Union Colony was the first planned Anglo-European agricultural settlement on the Piedmont, farming in the region was not new. During the late 1700s, bands of Cheyennes, Sioux, and Arapahos inhabited the Piedmont region. Among those groups, only Cheyennes practiced settled agriculture. However, like their Sioux and Arapaho neighbors, they abandoned horticulture in favor of a horse-based nomadism at the end of the eighteenth century.[8] As Anglo-European traders established forts along the South Platte River and some of its Front Range tributaries, they planted fields and gardens to supply their temporal needs. A handful of others, including Elbridge Gerry, settled on the Piedmont during the 1840s. Gerry, who claimed to be the grandson of a signer of the Declaration of Independence and who trapped furs in the Rockies during the 1830s, married two Lakota Sioux women and maintained a ranch and trading outpost at the confluence of Crow Creek and the South Platte River, just ten miles southeast of the future Union Colony. Gerry was ideally positioned to supply the waves of people about to descend on the Piedmont.[9]

For settlers moving west, the Colorado Gold Rush of 1858 foretold opportunity on the Piedmont. Gold strikes in 1858 and 1859 brought as many as one hundred thousand prospectors to the Rocky Mountain foothills west of the burgeoning city of Denver. Mining appetites fed agricultural settlements on the Piedmont, and this small slice of the region Americans commonly referred to as the Great American Desert, slowly filled in. During the 1860s, most of these new settlers claimed farms that stretched along both sides of the South Platte River as it extended to the north and east of Denver, enabling them to supply the new city and nearby miners with essential foodstuffs at prices inflated by the region's relative seclusion, growing population, and paucity of supplies. In 1863, Platte Valley farmers produced two million bushels of grain. Five years later, reporters for the Denver-based *Rocky Mountain News* made trips along the South Platte River and along some of its tributaries—Cherry, Boulder, and St. Vrain Creeks, as well as the Big and Little Thompson Rivers—describing an agricultural landscape filled with oats, wheat, corn, barley, and potatoes. Though settlement continued to spread through the well-watered areas of the Piedmont, agricultural growth remained modest in the decade following the gold rush. In 1870 the entire Colorado

Territory boasted 1,728 farms, the majority of which were located on the Piedmont.[10]

Those farms were the harbingers of vast ecological changes. Prior to Anglo settlement, the region played host to a complex set of interdependent ecological relationships above and below the land's surface. Buffalo grasses were the most prodigious plant species on the Piedmont, while wheatgrasses were common in the bottomlands adjacent to streams. Interspersed with these grasses were woody plants such as purple coneflowers, chokecherries, and sumac, some of which were staples within the complex of foods and medicines used by Cheyennes and other Indians of the region.[11] Groves of trees populated intermittent points adjacent to Piedmont streams. While settlers from the well-watered East viewed the Piedmont shortgrass prairie as sparse and plain, this belied a dense and complex web of life below the surface. Tiny fungi called *mycorrhizae* attached themselves to buffalo- and wheat-grass roots, functioning to supply phosphorous to individual plants and transfer that critical element to the larger plant community. These essential fungi were the workhorses holding together the flourishing plant life of an ecological community.[12]

Though the shortgrass prairie that characterized the Piedmont was a product of evolutionary plant, animal, and climate relationships, it was not a static ecosystem. Heavy rains and drought could alternately drown or kill the life-giving fungi below the surface. By contrast, periodic prairie fires increased the spread of helpful fungi. Some of those fires were ignited by Indians who embraced its role in supporting life-giving plants as well as how it could be harnessed to direct bison toward hunting parties. Bison grazing patterns also shaped this diverse shortgrass biome, temporarily reducing native grasses in some areas, effectively making space for woody plants to flourish.[13]

Human and climatological factors accelerated change in the decades prior to largescale settlement. Some of those were set in motion when Elbridge Gerry and others arrived in the 1840s. Bison were already disappearing from the Front Range region of the Plains that hugged the eastern extent of the Rocky Mountains. In fact, bison numbers had thinned through most of the Great Plains. In what historian Elliott West calls "a masterpiece of adaptive fluency," Indian nations such as the Comanches, Cheyennes, Sioux, Arapahos, Kiowas, and Plains Apaches adopted a horse-mounted nomadism to hunt bison to take advantage of Anglo-European market demands for buffalo hides.[14] Subtle climate changes may have already been impacting the shortgrass prairie as well. A period that climate scientists refer to as the Little Ice Age, which had kept earth temperatures below historic norms since the 1300s, was fading during the mid-nineteenth century, resulting in erratic weather patterns that disrupted the growth of life-giving fungi in Piedmont soils and altered historic bison migration patterns.[15]

It is difficult however to untangle the end of the Little Ice Age from alterations wrought by the growing stream of settlers, travelers, and fortune seekers descending on the Piedmont during the same period. The small number of farms and forts established to support trade and travel prior to the Colorado Gold Rush in 1858 enabled humans and livestock to trample existing grasslands. En route, they cut down much of the region's sparse supply of timber. Once the gold rush commenced, their impacts multiplied. Additionally, the growing number of farmers who concentrated along the bottomlands plowed through grasslands and cut irrigation ditches, severing interdependent plant communities that extended far beyond property lines. Viewing Indians and prescribed fires as primitive, settlers sought to remove both from the landscape. As a result, they eliminated processes and peoples whose management practices were adapted to the region. Ironically, Anglo settlers may have felt vindicated in these beliefs by the land's persistent productivity, since the phosphorous stored in Piedmont lands—supported in part by Indian land management—propelled plant growth. By 1870, when Union colonists arrived, the fragmentation of the Piedmont shortgrass prairie was well underway. While the new settlers imagined a landscape overflowing with market grains and produce, they had not yet concocted an agroecology to replace what they were dismantling.[16]

Settling the Union Colony

Horace Greeley and Nathan Meeker supplied a founding vision for the Union Colony intended to govern its early economic, social, and organizational development. Both were heavily influenced by the French socialist Charles Fourier and his American protégé, Albert Brisbane, who, during the 1830s and 1840s, inspired several cooperative agricultural settlements. Fourier believed that his communities, which he called industrial associations, should cooperatively own and cultivate a common set of lands. He emphasized that associations must adopt the most efficient machinery and up-to-date agricultural science to secure abundance. Fourier claimed that modern farming methods and cooperative enterprises would yield larger harvests on fewer acres and with less toil than existing practices. He and Brisbane believed that communitarian efficiency supported an equitable distribution of wealth while generating time for the kind of intellectual and cultural pursuits that enabled industrial associations to thrive. While Greeley and Meeker supported Fourier's ideas, they stopped short of complete communalism, believing that farms should be organized around the family and that property should be privately held. The associations that Fourier envisioned would be attained through collective planning and shared financing of community projects.[17]

After observing the failure of several Fourierite communities in Ohio and New York during the 1840s, Greeley concluded that the American West was the ideal region to see his ideas put into practice. He had little regard for Indians or their claims to the land, viewing the region as one vast public domain where planned Anglo settlements could flourish unobstructed by land grabbers and wealthy speculators. Having toured the West in 1859, Greeley recognized that the arid, sparsely populated region required irrigation for successful farming. In addition, he understood that building irrigation infrastructure depended on capital beyond the financial reach of individual homesteaders, necessitating western settlers to pool their resources to divert water to their crops. Thus, Greeley deduced that the same cooperation required to divert water could engineer thriving agricultural colonies. Further, Greeley argued that irrigation, when combined with efficient cultivation of crops through modern scientific methods, would enable farmers to produce bountiful harvests on small plots. Thus, colonists could live in close proximity, succeed in market agriculture, and work together to build community without the geographic distance created by farming on large acreages.[18]

Although Greeley's ideas about farming in the West were essential for establishing the Union Colony, he neither settled there nor contributed much in the way of physical or organizational labor. In that regard, Nathan Meeker played the largest part. A journalist and an idealist drawn to transcendentalist literature and utopian communities, Meeker befriended Greeley in the early 1840s after reading a series of articles about Fourier's industrial associations in the *New York Tribune*. Meeker then lived in a Fourierite colony in Ohio for three years. After the community collapsed and Meeker's first book—supported financially by Greeley—sold miserably, Meeker wrote a series of agricultural articles that Greeley published in the *Tribune*. In 1861 Greeley hired Meeker as a Civil War correspondent. Following the war, Greeley kept Meeker on as his agricultural editor. In 1869 Greeley assigned Meeker to write a series of articles on the Utah and Colorado Territories. It was there that Meeker became convinced that a cooperative agricultural colony in Colorado could thrive. Upon his return, he approached Greeley with his ideas. The aging *Tribune* editor responded enthusiastically: "I think it will be a great success, and if I could, I would go myself." Eventually, this support took the form of strategic planning, monetary aid, and generous editorial space in the *Tribune*.[19]

Following his meeting with Greeley, Meeker placed an ad in the *Tribune* giving a physical description of the colony, the philosophy it would embody, and the requirements for those who wished to join. Though Meeker had not chosen an exact location, he was persuaded by Denver newspaperman and real estate booster William Byers to consider the South Platte drainage. Meeker obliged, partly because of its proximity to the new Denver Pacific Railroad and a soon-to-be-built spur line of the Kansas Pacific Railroad. The text of Meeker's editorial highlighted the healthfulness of the land and its resources. He gushed that it was well watered, with rich soil, pine groves, and access to stone and other building materials, as well as coal for heating. Meeker also emphasized colony roads and rail access. He further argued that the mild climate and healthful land could support both cash crops and livestock, while still offering access to plentiful wild game.[20] Meeker punctuated his evocative—and somewhat exaggerated—description by stressing the region's unparalleled scenery.[21]

For the Union Colony to thrive, Meeker highlighted several requirements. First, the land would be purchased as a block to avoid land speculation, and membership and initiation fees were required up front. These funds would be used to form a town and develop irrigation. Farming acreage would be located in close proximity to the town and kept small—initially he proposed forty- to eighty-acre parcels—since he argued that modest irrigated plots would yield abundant produce and a tight-knit community. Meeker stressed the need to rapidly build up town infrastructure by enlisting those with skills in retail trades such as baking, blacksmithing, and milling. Finally, in a nod to the reform philosophy out of which many colonies originated, Meeker stated that temperance would be the rule of the colony.[22]

Meeker wasted no time organizing. At the initial public meeting in December 1869, fifty-nine people paid five dollar initiation fees to join the Union Colony and many more pledged to do so. Prospective settlers then formed an executive committee to oversee the new colony, electing Meeker president at its initial meeting and selecting Horace Greeley as treasurer. In January 1870, a locating committee, which included Meeker and other members of the executive committee, set out to select a site. In April they chose a valley on the Cache la Poudre River, three miles above its confluence with the South Platte River. Within a week, the executive committee agreed to purchase twelve thousand acres of primarily Denver Pacific Railroad land. In the meantime, four hundred applicants pledged at least $160 to obtain a Union Colony lot. Though Meeker undoubtedly played the largest role in founding the colony, he refused the suggestion that it be named after him. The executive committee then put forth the name of the *New York Tribune*'s chief editor. Thereafter, the Union Colony would be called Greeley.[23]

By some measures the town of Greeley experienced great initial success. Within one year of its founding, colonists built 232 houses for the more than one thousand inhabitants who had relocated there, making it the fourth largest town in the Colorado Territory. Colonists completed nine miles of irrigation canal and were constructing thirty additional miles to redirect the waters of the Cache la Poudre River to farmers' thirsty lands. Simultaneously, colonists finished churches and a school and were hard at work building a town hall. They also undertook cooperative efforts to build both grist- and saw-mills.[24] On initial glance, Greeley was thriving and owed much of its success to the ideals and efforts of its founders.

Despite this, Greeley settlers had reasonable complaints. For starters, Nathan Meeker and the Union Colony Executive Committee oversold the community. While several miles of Ditch Number Three were completed when settlers arrived in the summer of 1870, that water did little more than provide for in-town needs. Moreover, the cost of building additional ditches ran over budget and behind schedule. Consequently, the executive committee assessed additional fees on colonists to pay for canal construction. So, when Meeker and colony leaders presented prospective settlers with optimistic estimates regarding quantities of wheat, corn, vegetables, and potatoes that could be grown, actual residents viewed such claims with cynicism. For, without sufficient water, none of that was possible. Mary Norcross Tuckerman, an original Union colonist, cites several "indignant meetings" at which new residents complained about poor city lots and overly "rugged conditions." In addition, drought and poor hygiene in 1871 likely contributed to a typhoid outbreak that killed forty people. Then, in the summer of 1871, during the first full year of agriculture in Greeley, grasshoppers decimated local crops. These invasions recurred annually for at least three years following settlement. While Greeleyites planted a variety of crops, the sudden cultivation density made the community easy prey for insects.[25]

As parched settlers sought effective methods for killing grasshoppers and attempted to construct water delivery systems that could keep pace with population expansion, they required other food

Fig. 3. Greeley Colorado, 1870. Greeley Ditch #3, which supplied the town with water, is on the left. City of Greeley Museums. Used by permission.

sources. Ironically, the most nutrient-dense sustenance available to them was found in bison, an animal that settlers were complicit in annihilating. The railroads that Greeleyites courted in the 1870s to transport their future harvests killed millions of bison to clear the path for settlers and cattle who would replace them. So, when settlers dispatched hunting parties onto the plains to kill bison, they hoped that a once abundant animal would hold out until they could engineer sufficient ditches and crops to replace them.[26]

Irrigation and Speculation

Though Nathan Meeker emphasized the symbiotic relationship between founding economic and cooperative visions, Greeley's early history demonstrates that cooperation was merely a social means to an economic end. This was evident from the start. Colonists moving to Greeley understood that the town's success would depend first and foremost on agricultural productivity. Since plots were small, success required intensive cultivation of valuable cash crops. Further, they expected that abundant irrigation would propel impressive yields of potatoes, wheat, corn, beets, and fruit—all of which commanded high prices. In addition, they believed that easy rail access, a growing regional population—especially in Denver—and the area's booming mining industry would offer a steady market of customers for their produce. Moreover, many residents joined the colony in the hopes of making quicker profits than were otherwise possible through home-steading on the public domain, which would have entitled them to more acreage at less cost but promised neither water rights nor prox-imity to markets. The most valuable agricultural lands—such as those purchased by the Union Colony—were held by the railroads. Early tensions illustrated the primacy of Greeley's economic philosophy over its social one.

No issue demonstrated the supremacy of economic motivations more than speculation. Water was liquid gold. Once irrigation water flowed onto settler properties, land values instantly rose. By the sum-mer of 1871, Greeley residents had constructed twenty-seven miles of Greeley's Canal Number Two, enough to water forty thousand acres. According to the *Greeley Tribune*, the town's fledgling newspaper—edited by Nathan Meeker—this had the effect of more than doubling land values.[27] One colonist, F. E. Baker, originally settled on eighty acres of land five miles from town. Feeling isolated and concerned that irrigation water was not forthcoming quickly enough, Baker offered to sell the land and water rights to a Greeley merchant named Bill Dickens for one hundred dollars. Dickens declined, arguing that water would never flow onto Baker's property. Just one month later, Dickens was proved wrong, and it would cost him. Dickens agreed to buy the land and water right for $150. Baker regretted the move for, one year later, Dickens sold the same property and its water right for $500.[28] Water's economic power encouraged some Greeleyites to pocket their profits and leave. At the same time, others gambled on the trend by purchasing more land and water. Still others subdivided their holdings, despite taxes levied to discourage the practice.[29] Water manipulation made the Union Colony even as it unraveled its communitarian ethos.

With irrigation playing such a pivotal role, the process by which water diversion unfolded in Greeley was destined to further strain colony ideals. Initially, colonists planned four irrigation ditches to provide the town with water. Since Ditch Number Three diverted water for the town—and some adjacent fields—colonists commenced building it immediately. The far more expansive Ditch Number Two would channel enough water to farm crops for each of the remaining colonists. The Union Colony Executive Committee estimated that total costs for these two diversions would be $20,000. Unfortunately, Ditch Number Three alone required $25,000 to complete. With no funds available to construct a waterway for farmers, the Union Colony Executive Committee levied additional taxes on new lots and the subdivision of property to pay for it. Additionally, they redirected funds allocated for town construction to irrigation. This was a source of tension between settlers whose livelihood was based within the town of Greeley and sought its development, and those whose source of income was farming and who required irrigation for success.[30]

Financing irrigation projects was not the only sticking point, as diverting water from stream to farm proved more complex than anticipated. Poorly cut ditches eroded banks and seeped water, while canals without a consistent fall line caused water to pool up in one location while moving too swiftly in others. Water absorbed into ditch banks facilitated unwanted growth that sucked up water intended for crops. When that growth eventually died, it was carried downstream, resulting in ditch-clogging debris. In their cost calculations, Union Colony founders assumed incorrectly that maintaining ditches would be inexpensive. Colonists not only lacked the funds for irrigation, but the know-how to carry it through.[31]

Population growth in communities who diverted water from the same streams as Greeley also complicated founding ideals. This was especially evident during an 1874 conflict. Less than two years after Union colonists arrived, another group of settlers organized the town of Fort Collins. Situated along the banks of the Cache la Poudre River upstream of Greeley and organized by several of Greeley's founding settlers, Fort Collins also tied its fortunes to irrigated agriculture. Ignoring Greeley's prior water claims, Fort Collins initially diverted water with impunity. In the drought year of 1874, this created tension, as Fort Collins farmers drained most of the river's flow before it reached Greeley's main canal. Amid threats of armed conflict, Fort Collins farmers agreed to release some of the river's flow. Heavy late-summer rains dampened further conflict. Prior to 1874, Greeleyites trusted that their earlier water claims would protect them from upstream rivals. As a result of the conflict with Fort Collins, they pushed for laws that recognized their earlier water rights. Though this right of prior appropriation had been established by legal precedent in the Colorado Territory before 1874, it was not yet enshrined in its statutes. Ironically, Fort Collins irrigators soon concurred with their Greeley neighbors once newer settlers arrived and sought to draw river water even further upstream. Piedmont communities quickly learned that cooperating to gain sufficient water trumped gambling on corralling an abundance of this slippery resource.[32]

It would be a mistake, however, to view the water war between Greeley and Fort Collins solely as part of a larger learning curve on how to irrigate lands in the American West through cooperation. As historian Donald Worster has rightly observed, Greeley was founded on a core value in nineteenth century capitalism, namely that nature existed to be exploited for financial gain. While Nathan Meeker and Horace Greeley hoped that the mutual need for water would forge community, accumulation and speculation proved to be the stronger impulses. Once engineered, irrigation was capital, and the competition to appropriate the Cache la Poudre River was also a play to take the largest possible share of the region's wealth. If Greeley's claims were based on arriving first, then Fort Collins' water grab was rooted in geography. Being familiar with the location and ethos of Greeley's claims, they simply diverted water further upstream, enabling them to draw water earlier than their downstream cousins. This was bound to be a temporary gambit since developing precedents favored first claimants over those diverting from higher ground. Eventually, however, every Piedmont settlers would have to yield to laws and limits on what Worster calls "an economic culture of capitalism." Water was volatile. Consequently, settlers set about rationalizing nature to ensure a secure supply.[33]

Greeley's 1874 water war with Fort Collins provides just one example of what Piedmont settlers learned, inscribed into law, and then passed on to others. Greeleyites achieved success carving their senior water rights into law in 1879, when Colorado, which achieved statehood in 1876, created water districts based around the state's various watersheds, subsequently appointing a state engineer in 1881 to oversee measurement and distribution. New laws ordered districts to clarify the seniority and quantity of every water claim. Water users who were "first in time" based on their water claims, would be "first in right" to take water annually. The law also confirmed that a water right was guaranteed in perpetuity provided that the user could make "beneficial use" of it.[34] This Doctrine of Prior Appropriation was supported by a related set of laws and court rulings that enabled water users to cross private property boundaries to maintain their canals and ditches, move water across district lines and into other watersheds, and prevent excessive charges for water delivery.[35]

While Greeleyites succeeded in protecting their water rights, they did not manage them according to founding principles. Diverting water from its natural course was an expensive endeavor and possession of a water right meant nothing without the means to engineer its use. Moreover, Colorado's beneficial use statute provided no incentive for leaving water in the stream. Without the physical and financial means to irrigate, water rights could be forfeited. So, most Piedmont water users vested their water rights in mutual irrigation companies. Most of these companies formed when users pooled their resources and water rights to purchase and/or build irrigation works. Mutual irrigation companies—also called ditch companies—functioned as cooperatives. Farmers pooled their water rights in the company and were allotted voting power on company decisions in proportion to the quantity of water held. Typically, mutual irrigation companies were responsible for building and maintaining reservoirs and larger canals, while individual irrigators built and maintained the ditches coursing through their own properties. Mutual irrigation companies arose and functioned out of shared needs and the desire to build and protect their users' water wealth.[36]

In just over a decade following Greeley's founding, most of its water was managed by mutual irrigation companies whose members stretched out beyond the boundaries of the original Union Colony settlement. In 1877 the town gladly sold its corporate ownership of Ditch Number Two to the farmers who used its water, who then formed the Cache la Poudre Irrigation Company. In succeeding years, these farmers expanded the capacity of the ditch, eventually serving agricultural interests that were not tied to the original Union Colony. Ditches Number One and Four were eventually built, though not by the town of Greeley. Number One became the Larimer and Weld Ditch Company's property, while Number Four, which took water from the Big Thompson River beginning in 1881, became the Loveland and Greeley Canal, serving farmers on the outskirts of Greeley.[37] Nathan Meeker and Horace Greeley had once argued that cooperative irrigation and intensive farming would foster community and financial success. The first ten years of the Union Colony proved that the buying and selling of water drained most cooperative impulses dry.

Since Greeley was the first Piedmont colony, its experiences offered a template for those that would follow. Within a little more than a year of the town's founding, five other colonies sprang up in the region. These included Longmont, Evans, and Platteville. According to historian James Willard, there were over 2,700 colonists living in the region in May 1871. Each colony bought land as a block, primarily from the railroads, then sold memberships entitling colonists to land and water rights, prioritized irrigated farming, and promoted various cooperative ventures. And, like Greeley, each colony eventually abandoned the trappings of communalism but sustained efforts to grow cash crops on small, irrigated acreage. Though other Piedmont towns, such as Fort Collins, Sterling, and Brush did not copy Greeley's

cooperative philosophy, they mimicked its focus on water development and intensive farming.[38]

It is difficult to overstate just how thoroughly Piedmont settlers
developed their water resources in a short period following settlement. During the last two decades of the nineteenth century Colorado
alternated with California as the state with the greatest number of
total irrigated acres, with the Piedmont having easily developed the
greatest share of the state's water resources. This is remarkable when
one considers that California possessed greater agricultural wealth, a
longer settlement period, a climate that supported a far wider range
of crops, and more available water. By 1880 water shortages were
common in the South Platte watershed, farmers battled for their water
rights in courts, and more than a dozen mutual irrigation companies
scrambled to build storage reservoirs. Within a few short decades
following Anglo settlement, the Piedmont was a land remade by the
appropriation and engineering of its rivers.[39]

Dreams of a well-watered agriculture had material consequences on
the Piedmont landscape and beyond. Canals and ditches cut across
the shortgrass prairies, severing ecological communities that had
existed for millennia. Constructed reservoirs, whether dug into soil
or corralled into natural depressions in land, sunk permanent bodies
of water into areas accustomed to receiving less than fifteen inches
of annual water. This was no small presence. According to Elwood
Mead, Colorado's first professional irrigation engineer and future head
of the Bureau of Reclamation, from 1924 to 1936, there were at least
fifty small irrigation reservoirs on the Cache la Poudre River alone,
in 1903.[40] Water conveyances reached toward and extended from
each reservoir and became part of the complex and interconnected
network serving Piedmont fields which, until recently, had been a
contiguous collection of plains grasses. By fragmenting the Piedmont
ecosystem, engineered irrigation systems accelerated the decline of the
native shortgrass prairie, supporting its replacement with an exotic
agroecology. And the power to engineer the transformation existed
within the collective ability of mutual irrigation companies to own
and manipulate Piedmont streams.

Nature's Capital: Grass and Cattle

How does nature become capital? On the Piedmont during the late nineteenth century, water diversion transformed nature into a commodity. Water was monetized with the promise of irrigation and its value grew exponentially with the engineering of each new ditch and canal. This is where Nathan Meeker and Horace Greeley made fatal miscalculations. While they viewed water manipulation as a tool for building cooperative communities, settlers more commonly understood it as liquid capital.[41] But water was not the region's only wealth-building commodity. Just beyond the Piedmont's irrigated boundaries, entrepreneurs introduced cattle onto the Great Plains by the millions, believing that its shortgrass prairies, situated on the public domain, would provide free and nutritious feed in perpetuity. For the most part, these cattle empires of the 1870s and 1880s occupied a dryland geography outside the Piedmont purview. Yet, cattle needed water and it was in this context that cattle ranching and crop cultivation initially came together.[42]

The origins of open range cattle ranching on the Piedmont coincided with the Colorado Gold Rush of 1858, when miners and implement dealers entered the territory trailing cattle. When William Green Russell and his party struck gold near the modern city of Denver in 1858, prospectors seeking riches soon followed. Retailers trailed behind, hoping to make a fortune supplying miners and settlers. In addition to all manner of tools, newcomers required oxen to move their supplies and, eventually, to plow their fields. Some of those who moved cattle westward found themselves in Colorado with winter approaching and no suitable shelter for their animals. One such person was Jack Henderson. In the fall of 1858, Henderson was freighting supplies westward from Lawrence, Kansas. He turned his branded cattle loose during the winter and returned the following spring to find them not only alive but in better condition than he had left them.[43]

Henderson discovered something that Plains Indians had long known. Bison thrived on the same shortgrass prairie that Henderson's cattle did. Both animals are members of the bovine family and possess an internal organ called a rumen, which functions as a vast fermentation tank that enables them to convert grass—a low grade carbohydrate—into usable protein. Though plains settlers may not have been able to explain the evolutionary commonalities between bison and cattle, they understood its practical implications. The vast, imposing prairies invited the enterprising cattle rancher to strike it rich.[44]

Yet the biological similarities between bison and cattle were limited, as entrepreneurial ranchers learned. First, cattle require more grass than bison to produce similar weight gain. Bison consumed in one year what cattle on the same range ate in a five-month period. Additionally, bison evolved to accommodate climatic factors more effectively than imported cattle. During mild winters, cattle continued to graze, since snow totals in the region are low and snow continues to melt even during the colder months. However, more severe winters presented survival issues. Bison possess a greater degree of insulation than cattle and evolved in temperatures of forty degrees below zero Fahrenheit. Domesticated cattle can die when the thermometer dips below zero. Further, the large head and low profile of bison enable them to find forage through ice and deep snow, whereas cattle, with smaller heads and a taller stance, can perish under the same conditions. Amid their blind optimism, fortune-seeking cattle ranchers from the 1860s through the 1880s were slow to recognize these limitations. Eventually, it would cost them.[45]

No one personified the entrepreneurial optimism of the Colorado open range cattle industry more than John Wesley Iliff. Born in 1831 to a Methodist minister, Iliff grew up raising cattle on an Ohio farm. In 1859 he joined the throngs of gold seekers heading to Colorado. After achieving some retail success selling supplies to miners in Denver, Iliff invested money in land and cattle. Initially, he purchased weak animals at low prices from cattle drovers, turning them loose to graze on the shortgrass prairie. Iliff then sold fattened individual steers after two to four years, making a tidy profit. By 1870, as Union colonists initially settled their fledgling community, Iliff's cattle empire already occupied vast stretches of Northern Colorado.[46]

While there is no doubt that Iliff possessed great business acumen, timing and ecology boosted his success. When Iliff launched his cat-

Fig. 4. Portrait of John Wesley Iliff. History Colorado, Denver, Colorado. Used by permission.

tle business in 1861, vast resources were at his disposal. With bison in rapid decline and the federal removal of Indians, there was little competition for access to the shortgrass prairie that made up most of Eastern Colorado.[47] Further, in addition to a ready market of beef eaters in the growing city of Denver and nearby mines, Iliff signed contracts with the federal government to provide beef annuities to Indians living on reservations. Additionally, Iliff benefited from loopholes in the Homestead Act of 1862. The legislation enabled settlers in the American West to obtain 160 acres of the public domain for free if they were able to improve that land within five years. Though settlers were prohibited from obtaining additional land, Iliff contracted with his employees to file for holdings and then transfer title to him after five years. Through this method, Iliff was able to obtain fifteen thousand acres adjacent to streams on the Piedmont and in Wyoming. Water access gave Iliff significant power in Colorado ranching circles. Since many of Iliff's peers possessed little or no land adjacent to a stream, they depended on Iliff's benevolence for access—a benevolence he generally bestowed. Thus, while Iliff owned only fifteen thousand acres, he could rightly claim that adjacent ranchers who owned a combined six hundred thousand acres depended on his generosity to water their cattle.[48]

For other potential cattle entrepreneurs, Iliff's success provided a tempting template for their own endeavors. Though cattle prices varied from year to year, during the 1860s and 1870s they were a fraction of the money paid for fattened cattle freighted through the Midwest to terminal markets such as Chicago. During a single two-year period in the 1870s, Iliff purchased one thousand calves from a Texas drover at seven dollars per head. After Iliff's cowboys branded the cattle, they turned the calves loose to mature and fatten on the shortgrass prairie, occasionally rounding them up onto one of Iliff's ranches. Historian Agnes Wright Spring estimates that, for Iliff, the annual maintenance costs per head averaged less than one dollar.[49] After two years, these seven dollar cattle were rounded up one final time and shipped on the Union Pacific and Kansas Pacific Railroads to slaughterhouses in Kansas City and Chicago for a purchase price of approximately twenty-eight dollars per head. Though this represented a tidy profit, it was not uncommon. Thus, Iliff could argue without hyperbole that "this [the cattle business] is the grandest opportunity for investment ever offered; there are no uncertain risks attached to the business; the losses are almost nothing, and the profits many times those afforded by other investments." By the time John Wesley Iliff died in 1878, he had amassed a fortune worth over $1.1 million (over $30 million in 2021 dollars), while providing an example that would inspire hordes of would-be cattle kings hoping to mine the riches of the Colorado plains.[50]

The momentum created by the success of Iliff and others, as well as the heady promotional arguments of several opportunists, contributed to a cattle boom and bust on the Colorado Piedmont and throughout the Great Plains from 1880 to 1887.[51] During that period, over one hundred eighty corporations, with assets totaling over $75 million, sunk money into open range ranching on the Colorado plains. Initially encouraged by high market prices, most continued to invest even after market reverses in 1885 and a devastating winter in 1885–

86. After a drought-plagued summer the following year, the plains states, including Colorado, endured a brutal winter. Several blizzards alternately created layers of snow and ice, resulting in treacherous conditions for cattle that sought sound footing and the ability to forage through the snows to find grasses below. Instead, the thick crusts of ice cut through their legs as they walked and tore at their noses as they tried to forage through the snow. Cattle drifted toward river bottoms seeking food and shelter. Already emaciated by drought and an overstocked range, they died there by the thousands. One observer claimed that a traveler could walk along the north bank of the South Platte River that winter from Greeley to Julesburg walking on nothing but the carcasses of dead cattle. Later called the Great Die-Up, it signaled the death knell for the industry.[52]

Historians have offered many salient explanations for the collapse of open range cattle ranching. While brutal winters and short-term market losses played a role, they obscure deeper causes. Some have emphasized overgrazing. Cattle trampled and chewed through native buffalo and blue grama grasses, severing a highly evolved and interconnected web of plant communities. These were replaced by less palatable woody plants and exotic weeds. Climate change may have played a role as well, since the late nineteenth century marked a volatile end to what climatologists call the Little Ice Age, a period marked by cycles of drought, floods, and wild temperature swings. In addition, the same railroads that carried cattle to market functioned to bring more settlers to the region. Their farms and irrigation canals bisected the open range and increased pressure on limited water sources. Most importantly however, investors and cattlemen became victims of their own free market philosophy. Viewing prairie grasses and cattle as wealth-building commodities, they milked the resource dry. Unlike their Piedmont neighbors, they sought to capitalize on nature without investing in the infrastructure to maintain what they would inevitably alter.[53]

Building a Sustainable Piedmont Agroecology

In this sense, the economic vision of Greeley and other colonies on the Northern Colorado Piedmont provided a more sustainable template for long-term viability. With an emphasis on irrigation, water rights, and the growth of cash crops for the market, these agriculturalists were ideally suited to efficiently provide commodities for a growing nation. Following the Great Die-up, successful cattle ranchers had to mimic some of the practices of crop farmers. This meant concentrating their operations on much smaller plots and watering and feeding the animals within them. Ranchers built barns to shelter animals and store feeds, fencing their properties with barbed wire to direct their movements. Rather than directing cattle to water, ranchers brought it to their animals. Consequently, securing water rights and building delivery conveyances became essential. The need to own land with water access thus convinced many ranchers to purchase plots with reasonable access to Piedmont streams. A resurgence of irrigation canal and reservoir construction on the Piedmont ensued in the 1890s and 1900s. As the nation's network of rails connected the Piedmont with the rest of the nation, ranchers responded by replacing longhorn cattle—whose tough and stringy beef fell out of favor with consumers—with animals bred for their quality meat. Moreover, as ranchers replaced the amorphous open range with well-defined cattle ranches, stock raisers increasingly cut hay and grew alfalfa to feed their animals. These practices placed ranchers and crop farmers into each other's orbits.[54]

Philip Boothroyd exemplified many of these Piedmont changes. Born and raised in England, Boothroyd immigrated to Larimer County in 1872, purchasing land near Loveland with the intention of building up a ranch for his family. During his lifetime, he served as a justice of the peace in Larimer County and was active in the Farmers Educational and Cooperative Union, which supported the activities of small farmers in Colorado who resisted the influence of corporate agriculture. Compared with the cattle barons of his time, Boothroyd's ranching activities were modest; however, the mundane recordings in his journal explain the transformations in the cattle industry that occurred from the 1870s through the 1890s.[55]

Boothroyd's 1872 journal entries reflect the experience of a settler building an agricultural life for himself in a new place. In April and

May he planted a variety of crops, including cherries plums, beans, beets, melons, and corn. In this, Boothroyd mirrored the activities of his contemporaries in Greeley and other fledgling settlements nearby, planting hardy crops capable of bringing a decent market price. As might be expected for a newcomer, Boothroyd invested a great deal of time and expense building fences and corrals for his animals. These included horses, sheep, and cattle. Regarding cattle, Boothroyd's entries are enlightening, partially for what they reveal, but also for what they leave out. In the spring of 1872, he references "looking for cows" that had wintered on the public domain. His only other reference to caring for cattle was a May 2 entry in which he records that Blanche had a calf. In the back of each year's journal, Boothroyd itemized his daily expenses. While there were numerous entries for food and supplies, there was only one expense in 1872 directly related to cattle. It was an entry of five dollars to feed cows. Boothroyd operated under much the same philosophy as the larger cattle ranchers in the area: namely, that most of what was necessary to feed and maintain a herd of cattle was free of charge and readily available on the shortgrass prairie.[56]

Boothroyd's 1890s journals portray a rancher paying much closer attention to managing his livestock. In the winter months he recorded feeding cattle and hauling manure daily. In May, stock feeding ended, Boothroyd branded his cattle, and then turned them out to "pasture" through the warmer months. In the meantime, he fertilized his garden and trees with livestock manure, irrigating almost daily. In fact, no subject received more attention through the summer months than irrigation. This included not only letting water flow to his crops but also "attend[ing] to ditch." Throughout the warmer months, Boothroyd emphasized cutting hay and alfalfa and storing it away. He also established a corn granary and purchased wheat and oats from neighbors to supplement his winter livestock feeding. In September, he began harvesting crops and storing them away for the colder months. Meanwhile he initiated purchasing cattle in September, a practice that continued into November. During this period, he commenced feeding cattle again—his herd more than doubled during the

Fig. 5. Advertisement for Philip Boothroyd's Waterdale Stock Ranch. Philip H. Boothroyd papers, COU:158, box 3, folder 7, Rare and Distinctive Collections, University of Colorado, Boulder Libraries. Used by permission.

1890s—a practice that received daily attention starting in December. Sometime during the 1880s, Boothroyd began a breeding program on his ranch, advertising Aberdeen-Angus bulls and steers for purchase. Between 1872 and the 1890s, maintenance of stock moved from a peripheral to a central part of Philip Boothroyd's life.[57]

Boothroyd's experiences exemplify the transformation of cattle ranching and, to a lesser extent, of crop farming on the Northern Colorado Piedmont by the turn of the century. Raising cattle required managing every aspect of the herd. It was critical for ranchers not only to acquire feed but also to raise much of it themselves. With smaller herds and tighter markets for cattle, it was essential to be up-to-date on the feeds that efficiently fattened these animals. In Boothroyd's case, his choice to cut hay, grow alfalfa and corn, and purchase additional hay and oats from other farmers also reflected the research recommendations of agriculturalists at the newly minted land grant college in Fort Collins.[58] By 1900 the day when herds of cattle could be turned out onto the plains and forgotten had faded.

Further, Boothroyd's attention to irrigation reflected the additional need for a water supply. The same irrigation ditches that supplied his crops also supplied his herd.[59] If successful ranchers needed to manage an animal's intake, then the same was true of its output. On a tightly managed stock farm, manure was composted and repurposed as fertilizer for crops. The economics of feeding extended to the quality of the meat as well. In this, Boothroyd's choice to select and market a particular breed of cattle known for its attractive features and tasty meat mirrored choices his peers made at the same time.

The paths that drew Piedmont area ranchers into the orbit of the region's farmers were traveled in both directions. Economic opportunity accounts for some of this, as Piedmont farmers viewed ranchers in need of feeds as customers for their crops. Yet the free market offers a narrow and unsatisfying lens for understanding the evolving relationship between crop cultivation and livestock ranching. Clarity requires tracing how agricultural science translated into a farm practice that brought crops, cattle, and water together. Applicable research in fields such as chemistry, microbiology, and hydrology worked its way into fields, pastures, and ditches. The literature that farmers read and disseminated translated into farm experiments and eventually into established practices. The practical application of agricultural sciences further intensified in 1887, when Colorado's new land grant college established an agricultural experiment station at Fort Collins.[60] An arm of CAC, the experiment station concentrated the majority of its early efforts on applying research conducted elsewhere to Piedmont farming.[61] Thus, viewing the region's transformation requires a dual lens that extends into the scientific literature beyond the Piedmont to comprehend evolving farm practices within it.

The Greeley settlers of 1870 were familiar with findings in the agricultural sciences during the previous fifty years. Nathan Meeker and Horace Greeley had been arguing that meticulous soil cultivation alongside the skilled deployment of science and technology would yield prosperous communities. Irrigation and escalating land values added urgency to this philosophy since they depended on sufficient capital investments applied to modest acreage. By 1870, agricultural scientists in Europe and the United States concluded that all plants required potassium, phosphorous, lime, magnesium, sulfuric acid, and nitrogen. Determining that most plants deplete these nutrients from the soil, they sought to quantify the loss for the more common cultivars and determine how they might be replenished. Since nitrogen was far and away the most critical of these elements, researchers sought to understand the method by which plants absorb nitrogen and how farmers could make it available to plants. They determined that decomposed animal manures and plant material emitted nitrogen in the form of ammonia. Agricultural chemists also demonstrated how lime, in its various forms, added nothing to plants but, when applied in controlled quantities, it regulated the release of soluble matter to ensure a more consistent uptake of nutrition. Further, researchers contributed to the growing practice of planting legumes, such as clover and peas, through experiments that showed that plowing legumes into the soil restored vital nitrogen.[62]

Nearly twenty years later, David Boyd, one of the original Union colonists, wrote a Greeley town biography that demonstrated how locals put the new agricultural science into practice. Boyd states that farmers in Greeley and its immediate environs learned to follow prescribed crop rotations that were adapted to specific soil types. While some early settlers had tempted fate by cropping the same grains in successive years, reduced yields quickly killed that practice. Farmers learned to cultivate wheat rotated with potatoes, oats, barley, cabbage, and corn and synchronized their practices to the soil types and water requirements of their specific plots of land. For example, Boyd points out that the light and sandy soils near Greeley attained a depth of up to fifty feet before they encountered heavy clay. These soils retained water and essential plant nutrients comparatively better than other lands in the region, resulting in high yields of potatoes and earning Greeley a national reputation for quality spuds. By contrast, farmers in this microclimate found that their soils suffered from nitrogen deficiencies. Hence, success required hauling manure to their fields. Farmers on these lands were the quickest to adapt horses and cattle to their farming regimen, in large part for the nitrogenous manure that resulted. In summarizing what Greeley farmers had learned in their first twenty years, Boyd echoed the sentiments of the agricultural scientists of the previous half century when he stated, "In nearly all cases the question can be narrowed down to this, 'Is there enough phosphoric acid, potash, and nitrogen in the soil to supply the yearly drain of those substances?'"[63]

Regarding crops that restored nitrogen to the soil, Piedmont farmers and ranchers responded almost immediately to scientific developments and evolving agricultural practice. Experiments completed in Europe during the 1830s and 1840s demonstrated that legumes, when plowed into the soil, boosted nitrogen levels and crop yields.[64] Alfalfa performed this task more efficiently than any other legume. Yet, while Greeley farmers employed legumes such as red clover and peas in their crop rotations in 1870, alfalfa did not appear in any quantity for another decade. Its appearance mapped onto its geographic spread in the latter half of the nineteenth century. Introduced to California from Chile in the early 1850s, alfalfa immediately boosted soil fertility, but its roots died in colder climates. As a result of selective breeding for cold hardiness, alfalfa made its way northward, eventually producing strains that could survive frigid Minnesota winters by 1900. Along its path of leguminous conquest, farmers first planted alfalfa in the South Platte Valley in the 1860s. It was a mainstay by 1880.[65] M. J. Hogarty, another pioneer Greeleyite, demonstrated why. In 1889 he grew a crop of alfalfa and a crop of clover, then tilled the crops into the soil and planted potatoes. Hogarty harvested doubled his yield on the alfalfa lands.[66]

Widespread adoption of alfalfa was also tied to livestock. As Piedmont farmers sought greater quantities of animal manure, they experimented with cut alfalfa and found that, when dried and cured, it provided a more nutrient-dense food source than hay or clover. Alfalfa's superior nutrition not only resulted in a manure rich in nitrogen but in livestock that gained weight more rapidly than on other feeds. This was no small matter as open range cattle ranching collapsed and ranchers who survived needed nutrient-dense feed for their stock. According to a study completed at CAC's Experiment Station in 1889, steers gained seven pounds of weight for every one hundred pounds of alfalfa consumed, making it the cheapest, most efficient, locally grown feed available. For ranchers such as Philip Boothroyd, who grew the crop and fed it to his angus steers, alfalfa was a cash cow.[67]

While the Piedmont was a relatively young agricultural region in 1900, it was modern by the standards of the day. From the first sizable agricultural settlement at Greeley in 1870, farmers in the region participated in a dialogue with chemists, hydrologists, microbiologists, and irrigation engineers. They embraced the new agricultural college, while reading and integrating findings within its publications. Farmers sought out extra-local markets for their crops and actively courted multiple rail lines for transporting their goods. The quick integration of alfalfa, alongside growing scientific evidence for the crop's value, demonstrates the rapid evolution of crop rotations and farm management techniques in the region. Further, by the turn of the twentieth century, 90 percent of improved agriculture on the Piedmont was irrigated, made possible by sophisticated engineering of canals, ditches, and storage reservoirs. Even the open range cattle ranching that dominated much of the region until the mid-1880s offers evidence for the modernization of Piedmont farming, as surviving ranchers adapted to the need to manage their stock and integrate their practices with the prevailing crop farming of the region. An aerial view of the Piedmont in 1900 would have revealed an extensive network of irrigation works coursing through clearly defined small farms, and several well laid out communities, all tied together by a functional transportation network.[68]

In 1900 power on the Piedmont was dispersed even as farmers embraced the free market and wealth accumulation. Land holdings remained relatively small and there was little evidence of corporate control, especially of the kind that was already present in California.[69] Further, farmers and ranchers retained autonomy over crop choice, producing the majority of their goods for the market and buying consumer goods from elsewhere. In addition, they employed little or no outside labor on their farms. Certainly, there was no permanent

laboring underclass.[70] Yet, small landholdings and farmer autonomy did not preclude embracing corporate agriculture. These were
not the mythic yeoman farmers celebrated by Jeffersonians for their
independence from outside influence. Land and water speculation
were common. Farmers' embrace of the agricultural sciences was
motivated far more by economic gain than any autonomy it might
afford them. In aggressively courting transportation networks for
their goods, farmers tied themselves to far-flung capital and markets. Piedmont farmers were opportunistic capitalists, cultivating a
diverse set of crops in a region characterized by broadly dispersed
land and wealth.

The imprint of Nathan Meeker and Horace Greeley remained on
the Piedmont in 1900, though it was difficult to locate. Even as the
town of Greeley ditched the Union Colony's communitarian philosophy, the region's farmers continued to emphasize irrigation and
the agricultural sciences. And though speculation and free market
capitalism ruled the day, the Piedmont remained a region of small,
owner-operated holdings that prioritized high value cash crops. Over
the next thirty years, parts of this Piedmont snapshot would remain
stubbornly resilient, even as a new industry sought dominance. As
for Horace Greeley and Nathan Meeker themselves, they did not
live to see the changes. In 1872 Greeley forged an unlikely run for
the presidency and was resoundingly defeated by the incumbent
Ulysses S. Grant. Distraught over the loss and reflecting on his perceived failures, Greeley was committed to a sanitarium and died
before the electoral votes were tallied. Nathan Meeker founded the
Greeley Tribune in late 1870, hoping to resume his career as a journalist. Though the newspaper survived, Meeker fell deeply into debt.
In 1877 he sought and gained a federal appointment as an Indian
agent on the White River Reservation in Western Colorado. His
heavy-handed attempts to turn Utes into yeoman farmers inspired
an uprising during which Meeker was killed in 1879. Six years later,
following the federal government's forced removal of Ute Indians,
Anglo settlers founded the town of Meeker on the site where Nathan
Meeker was killed.[71]

2 Capitalism and Sustainable Farming

When we try to pick out anything by itself, we find it hitched
to everything else in the Universe.
—JOHN MUIR

On November 21, 1901, three thousand visitors gathered in the small
town of Loveland to celebrate the completion of a new factory. It
included two railway presidents, socialites from Denver, and four hundred students and faculty from nearby Colorado Agricultural College
(CAC). The most notable attendees, however, were the directors of
the new Great Western Sugar Company. Three years earlier Charles
Boettcher, John Campion, and several associates pooled capital to
build a factory for refining sugar beets into sugar. After an initial
venture in Grand Junction, Colorado, failed in 1899, they were lured
in 1900 by farmers and local promoters to consider Loveland. Tests
on experimental plots of sugar beets in the vicinity revealed that the
region could grow beets that exceeded industry standards in yields
and sugar concentration. To sweeten the proposition, Loveland mer-
chants and farmers offered Boettcher and his colleagues 1,500 acres
of land, an $8,000 bonus, and a guarantee that farmers would plant
a minimum of 3,500 acres in beets during each of the next three
years. Planting began in the spring of 1901 and harvesting was well
underway when Loveland christened its beet sugar factory in November. Three-and-a-half months later Loveland's factory had processed
58,000 tons of beets into 139,000 bags of granulated sugar, netting a
profit of $75,000. It was a modest success that propelled rapid eco-
nomic, social, and agricultural transformations on the Piedmont.[1]

In 1900 most Piedmont farmers retained autonomy over crop cultivation from seed, to harvest, to sale. Over the next two decades, the entire region came to be dominated by a single crop: sugar beets. Most farmers planned their operations around this vegetable, contracting to sell their entire harvest to Great Western Sugar, a corporation that rapidly developed a monopoly over sugar refining in the region. By 1912 the Piedmont supplied more sugar than any other region in the United States. How did that happen and what human and agroecological consequences attended the transformation? This chapter answers those questions.

Despite an efficiency-driven market orientation prior to 1900, few Piedmont farmers possessed the capital, relational networks, or access to expertise that characterized the beet sugar industry. Sugar beets remade the Piedmont upon recognizably capitalist lines. The new crop commodified land and labor. Planting, cultivation, harvesting, and processing beets were each independent operations whose end product was sugar, a commodity that could be shipped anywhere and was indistinguishable—in look, color, taste, and chemistry—from the sugar produced elsewhere.[2] The cultivation and harvesting of sugar beets—the most time- and energy-intensive portions of sugar production—were performed by an elastic quantity of migrant hand laborers, calibrated to address the number of acres planted in beets and the anticipated harvest tonnage. Seeking to rationalize the work of making sugar, Great Western introduced a specialized division of labor that included seasonal workers in its sugar factories, researchers who aimed for more sugar per acre at lower production costs, field men who operated as company mouthpieces on farms, and employees in company-controlled subsidiary operations such as railroads and mining. Like its corporate peers, Great Western actively sought government subsidies by courting research at land grant colleges and the USDA and lobbying for industry-friendly policies. The sugar industry established a host of social and economic relationships characteristic of modern capitalism.[3]

Industrial agriculture comprises one critical subcategory of capitalist relationships in agriculture on the Piedmont. As historian Deborah Fitzgerald points out, the use of the term industrial to apply to agriculture emerged in American society during the 1910s and 1920s to contrast the efficiencies that had developed in urban factories with the perceived backwardness of the American farm. Critics argued that agriculture should be modeled along lines established in other industries. Practically speaking, this meant scrutinizing timelines for planting, cultivating, harvesting, and processing. Farmers should specialize in high production crops, adhere to narrow cultivation timelines and "de-skill" the labors of commodity production so that they could be performed cheaply by an interchangeable workforce. As with other industries, extra-local inputs were essential, especially if farmers were to make nature function according to the dictates of an industrial timeline. Local inputs were simply not sufficient to industrialize local landscapes. So, advocates of industrial agriculture, such as economists, bankers, government officials, and agro-industry, promoted the development and application of increasingly sophisticated machinery, fertilizers, and chemicals. According to historian Colin Duncan, industrial agriculture posited an inversely proportional relationship between agroecology and farm commodities. As society demanded a more diverse and complicated array of farm goods, successful production agriculture simplified complex ecological relationships to conform with market demands. By 1900 each of the components of industrial agriculture existed in discrete locations, such as the Red River Valley of the Dakotas and much of California. It developed unevenly prior to World War I, and then became a mantra for most farmers—particularly in the West—after the conflict.[4]

During the early twentieth century, Great Western Sugar and its growers embraced the trappings of capitalism and industrial agriculture even as they accepted and operated within the limits presented by the agroecology and economic arrangements in existence before the sugar company built a single factory.[5] Though Great Western monopolized Piedmont sugar refining, sustained beet cultivation depended on diverse crop rotations and meticulous attention to soil health in order to cycle the nutrients that beets extracted back into the earth. The company and its growers adopted harvest and refining practices that made beet byproducts available to feed limited numbers of livestock during winter months. In turn, their metabolic energies enabled farmers to enrich their fields with measured quantities of manure, a potent source of nitrogen and phosphorous, while marketing fattened livestock. Manured fields bolstered beet yields and sugar production while virtually eliminating the need for purchased fertilizers.[6] The need for balance in crop and livestock cultivation was also reflected in relationships between Great Western and its growers. While the sugar company possessed concentrated capital along with refining expertise and capacity, Piedmont small farmers owned most of the region's land and water rights. To obtain beets, the company needed farmers to sign beet-growing contracts and, since sugar refining is a water-intensive operation, it relied on the benevolence of its producer clients. Mastery of the Piedmont sugar beet landscape was evenly divided between the sugar company, its growers, and environmental factors that neither could ignore. Thus, corporate agriculture accelerated the growth of industrial capitalism on the Piedmont even as it held the existing agroecology in place.[7]

During the first three decades of the twentieth century, the Piedmont's value as the United States' largest sugar producer was built in equal parts on the global and the local.[8] The hand of economic protectionism, alongside consumer demand for sugar, and the global ambitions of the United States made the Piedmont beet sugar industry possible. Economies of scale and the specific needs of the beet sugar industry required a division of labor and expertise possible only within off-the-farm labs and labor markets. Yet, despite these markers of industrial capitalism, cultivation rested on maintaining an agriculture that had been solidified prior to Great Western Sugar's arrival. Sustainable crop rotations, livestock manure, and the health of Piedmont soils were inseparable from capitalism, industrial agriculture, and the global sugar landscape.

Sugar Politics and Economic Protection

The biological origins of corporate agriculture on the Piedmont are found in an odd-looking vegetable called the sugar beet. Off-white in color, shaped like an overgrown carrot, and possessing a taproot that extends several feet into the earth, this tuber provided the only concentrated source of sugar that could be grown and refined profitably outside of tropical climates at the turn of the twentieth century. Though adaptable to a variety of soils, sugar beets are best cultivated in well-worked ground in regions with abundant sunshine and an average temperature of seventy degrees Fahrenheit during prime growing months. They require frost-free days throughout those months and thrive when cool nights alternate with warm days. Since a successful beet crop needs a minimum of twenty-five inches of annual rain, semiarid regions, such as the Piedmont, require plentiful irrigation to grow them.[9]

Global changes in sugar consumption and processing during the nineteenth century help to explain the desirability of Piedmont beet cultivation. Between 1800 and 1890, worldwide production of sugar exploded from 245,000 tons to six million tons. How was this possible? During the nineteenth century, industrializing nations such as Great Britain and Spain developed cane sugar industries in their colonies and protectorates in tropical and subtropical regions of the world, including India, Cuba, Hawaii, and the Philippines. At the same time, continental European nations led by France, Germany, Austria, and Russia cultivated their own domestic beet sugar production, in part to maintain sugar independence from England's imperial possessions. As the quantity of sugar skyrocketed, manufacturing costs plummeted, propelled by factory innovations and efficiencies in seed breeding, cultivation, and harvesting. These factors made sugar accessible across socioeconomic classes. Moreover, sugar provided quick energy for workers' long hours during the Industrial Revolution and cheap calories for their impoverished families. Further, as urban families consumed more manufactured products, sugar became a universal sweetener in a host of industrial products such as jams, cakes, candies, and breads and an indispensable additive to drinks such as coffee and tea. By 1900 British consumers, who occupied the leading edge in this sweetness revolution, ingested 14 percent of their total caloric intake from sugar.[10]

Fig. 6. Cross section of sugar beets in the ground. Note the deep taproot. Beets demanded well-cultivated soil at planting and a great deal of energy to extract at harvest. Great Western Sugar Company Records, CSU Morgan Library Archives and Special Collections. Used by permission.

Many of these global factors that spurred sugar consumption and prompted imperial nations to cultivate and process their own sucrose were mirrored in the United States in the late nineteenth century. Between 1880 and 1900, U.S. sugar consumption rose from 1.3 to 2.4 million tons, making the United States the world's largest consumer.[11] Most of that sugar was imported from the cane fields of Cuba, where imperialism and plantation agriculture created oppressive conditions for workers. American industrialists and local beet sugar boosters argued that sugar beets could wash the stain of exploitation from the hands of consumers by enabling them to purchase a domestic sweetener cultivated by the hands of hard-working American farmers, while delivering a thriving new industry and healthy agricultural communities to the rural regions of the United States.[12] The German state was the model that beet sugar promoters most often highlighted since it offered scientific support and high tariffs to develop an industry that produced more sugar than all Caribbean cane fields combined, supplying its entire domestic demand. Germany's example of state-supported sugar independence offered a template for American industrialists, beet boosters, and the federal government. According to Charles Saylor, the USDA's primary sugar beet investigator in 1901, sugar independence would "eliminate the foreign grower as a factor in the supply of our daily wants, our business methods, and the emergencies of war," while it generated new opportunities for farmers and manufacturers.[13] Sugar beet promoters, entrepreneurs, and the USDA agreed that a thriving beet sugar industry was an unqualified national good requiring federal support.

It was not until the Republican-led Congress passed the Dingley Tariff in 1897, that the beet sugar industry took off. The tariff set rates on sugar imports at 1.685 cents per pound, which effectively doubled previous duties. Of equal importance was the fact that the rates applied to raw sugar as well as refined sugar. The American Sugar Refining Corporation (ASRC), commonly known as the Sugar Trust, imported minimally processed raw cane sugar from its tropical suppliers and refined it into the granulated sugar most Americans consumed. ASRC's control of as much as 95 percent of all refining capacity

	United States Production		Imported Cane Sugar	Consumption	
	Mainland (tons)	Colonies (tons)	(tons)	Total (tons)	Per Capita (pounds)
1881–1885	113,000		1,172,000	1,305,000	48
1886–1890	173,000		1,457,000	1,630,000	53
1891–1895	295,000		1,899,000	2,194,000	65
1896–1900	345,000	83,000	1,945,000	2,373,000	64
1901–1905	604,000	540,000	1,824,000	2,968,000	73
1906–1910	848,000	818,000	1,898,000	3,564,000	79
1911–1915	1,021,000	1,070,000	2,095,000	4,186,000	86

Fig. 7. U.S. sugar production and consumption, 1881–1915, five-year averages. Graphic created by Philip Riggs and adapted from Ballinger, *History of Sugar Marketing*.

in the United States during the 1890s enabled it to set retail prices and destroy competition. Consequently, when Congress debated tariff rates, Sugar Trust lobbyists sought to minimize or eliminate duties on raw sugar to maintain their stranglehold over the industry. Their failure in 1897 meant that they would be required to pay the full tariff duty, providing beet sugar industrialists, who were not beholden to the Sugar Trust, the economic space they needed to germinate.[14] The potency of the Dingley Tariff signaled the future political importance of economic protectionism for the beet sugar industry.[15]

The Piedmont beet sugar industry materialized during an era of high tariffs and notable imperial politics. U.S. victory in the Spanish-American War in 1898 placed Cuba, the Philippines, and Puerto Rico into American hands. While Puerto Rico grew little sugar cane, the Philippines possessed the land base and human resources to eventually become a significant producer. Since the Philippines and Puerto Rico remained U.S. colonies after the war, politicians divided over whether a tariff should be applied to goods produced there. No such limitations applied to Cuba since the prewar Teller Amendment guaranteed its official independence and initial public sympathy for the new nation, which exported almost all of its sugar to the United States, remained high.[16] Would the United States leave the fledgling nation to fend for itself by imposing full import duties on Cuba's most critical export?

The answers to those questions fused race with politics and economics. Sugar beet industrialists in the United States argued that protecting their industry supported a civilized, white, industrialized class of farmers who delivered a commodity that resulted from technological advances in seed breeding, chemistry, and cultivation. American beet sugar was emblematic of progress and ingenuity. Further, it supported a thriving class of farmers since sugar beet cultivation paid farmers well and generated high land values. Senators such as Francis Newlands went further on behalf of the American West, claiming that sugar beets would prime the pump of irrigation projects that were foundational to prosperous small farmers in the region. Contrasting the arid West with the humid tropics, beet sugar industrialists claimed that cane sugar's cultivation landscape and its workers were degraded and incapable of civilization, in cultural habits and in agricultural production. Some even went further, claiming that Anglo digestive systems were tainted by foreign cane sugar and that only high tariffs could support the "Aryan stomach." In this line of thinking, economic protection supported more than just domestic sugar production; it supported white ideas about civilization.[17]

Judicial and legislative decisions satisfied many beet sugar industry wishes. In 1901 the U.S. Supreme Court ruled, in *Downes v. Bidwell*, that Congress could impose tariff barriers on colonies, subjecting them to the same rates paid by foreign nations. Subsequently, Congress allowed Philippine raw sugar to enter the United States with a modest 25 percent reduction in duty. Two years later, Congress granted a 20 percent tariff reduction for Cuba, the United States' principal foreign supplier. Though beet sugar industrialists complained that preferred status for Cuba would cripple the industry, rapid growth and high returns during the first decade of the twentieth century suggested otherwise.[18]

Building the Piedmont Beet Sugar Industry

While economic protection was critical to the formation and success of the Piedmont beet sugar industry, a strong tariff did not guarantee success. Insufficient capital, limited crop expertise, and available labor all presented obstacles. Unlike the market crops cultivated by Piedmont farmers in the late nineteenth century, sugar beets required more processing to extract a usable product for the market. It was necessary to raise large sums of capital to build a factory and purchase machinery that could perform the physical and chemical energies required to transform an innocuous vegetable into sugar. Labor and expertise played roles as well. Investors needed sufficient sugar beets to process and a critical mass of farmers who would contract to grow a crop with which they had no experience. Further, cultivating and processing sugar beets required an army of seasonal field and factory workers. In a region with a negligible pool of available labor, sugar beet agriculture presented significant risks for farmers and investors alike.

Despite what the region lacked in capital, labor, and knowledge, the Piedmont did not lack for beet boosters prior to 1900. The experiment station at CAC was among the most prominent. During the 1890s, the experiment station conducted various studies that established that the soils and climate of the Piedmont could produce beets whose sugar content and sheer tonnage per acre would rival any region on earth. Experiment station publications explained that the Piedmont region also possessed abundant stocks of lime, potash, and coal, all necessary for processing sugar beets. Colorado officials also played critical roles in boosting sugar beets. In 1872 the territorial legislature (Colorado became a state in 1876) entertained a bill that would have paid a subsidy to any organization that built a factory capable of refining two hundred barrels of sugar per day. It failed by one vote, largely over financial concerns. Then, in 1899, the Denver Chamber of Commerce financed a competition that offered cash prizes totaling $1,000 for the best sugar beet crop grown "on a commercial scale." It was prominently advertised in several experiment station publications.[19] As with local boosters, the state legislature sought industries that could bolster the state's economy and attract new settlers.

Evidence that sugar beets were a viable crop, combined with a strong protective tariff, and efforts by promoters propelled industry growth. Once Great Western Sugar and its Loveland growers profitably cultivated and refined sugar beets in 1901, other Piedmont communities quickly followed. In the span of just over two years, from 1901 to 1903, five additional companies built Piedmont beet sugar processing facilities, using much the same template as the one established at Loveland. In Greeley, Longmont, Eaton, Fort Collins, and Windsor, local boosters attracted willing investors from outside their immediate vicinity to invest in a sugar factory. Eager capitalists were then willing to construct and operate a facility, once locals had secured multiyear contracts with farmers for a minimum number of acres planted in sugar beets (generally between 4,000 and 6,500 acres), a guaranteed supply of water for beet processing, the donation of a suitable site on which to build the factory, and a railroad right-of-way for transporting beets to factories and refined sugar to market.[20]

Great Western consolidated and grew in the early twentieth century at an astonishing rate. In 1901 Colorado's fledgling beet sugar industry caught the eye of the ASRC, headed by Henry Havemeyer, one of the richest men in America. Havemeyer controlled almost two-thirds of the sugar refining capacity in the United States. In 1901 sugar beets supplied 7 percent of the nation's domestic sugar supply. Fearing that upstart beet sugar companies could threaten his stranglehold on the industry, he employed several strategies to acquire each of the Piedmont's factories.[21]

Havemeyer's initial tactic was simply to purchase each factory. When owners refused to sell, he resorted to underselling Colorado beet sugar in its consumer markets. Since the plains and the Midwest provided outlets for Piedmont sugar companies, the ASRC flooded those regions with cane sugar at well below market value. Knowing that consumers would purchase the cheap cane sugar, many retailers refused to carry the Piedmont product since it was being undersold by an average of 20 percent. Factory operators, such as Charles Boettcher, countered by selling sugar in markets further east. However, he was aware that such a strategy would only work once and that Havemeyer would counter it the following year. Other factories attempted to store their sugar rather than sell it at below market rates. This also was a temporary strategy since none of the Piedmont operators had enough capital to hold onto their sugar for more than a season. Havemeyer put additional pressure on factories by filing articles of incorporation to build beet processing factories near existing ones. If built, Havemeyer would then entice growers to sell their beets to his factories at a better rate than they had received from the local one. This was more pressure than any of the sugar companies could bear, and by the end of 1903, all six factories—located at Fort Collins, Longmont, Greeley, Eaton, Windsor, and Loveland—sold a controlling interest in their companies to ASRC.[22]

Initially, Havemeyer was content to own companies that would otherwise have been competition. So, he appointed a Colorado native, Chester Morey, to oversee his financial interests there, while allowing each of the six factories to be managed independently. But problems

Development of the Sugar Beet Industry in Early Twentieth Century on the Colorado Piedmont

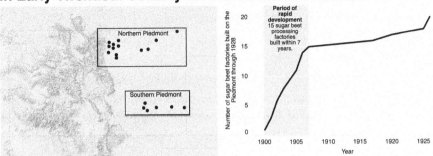

Northern Colorado Piedmont
Development of the sugar beet industry centered along the South Platte River and its tributaries.

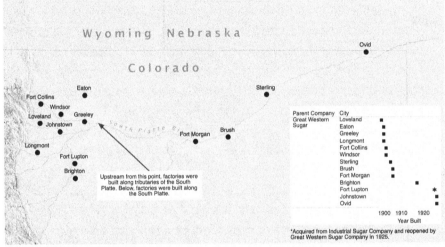

Southern Colorado Piedmont
Development of the sugar beet industry centered along the Arkansas River.

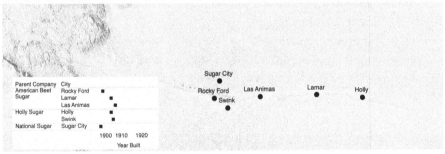

Fig. 8. Development of the beet sugar industry on the Northern and Southern Colorado Piedmont. The entire sugar beet harvest on the Northern Piedmont was refined by Great Western Sugar, while three companies refined beets on the Southern Piedmont. Map created by Philip Riggs.

with this arrangement mounted from the start. Growers clashed with management in 1903 over prices paid for beets, and the ongoing insecurity that there would not be enough laborers for cultivation and harvesting. Consequently, they formed the Sugar Beet Growers Association to press for more favorable contracts and a secure labor force. Factory management struggled with more than just growers. They relied heavily on the Colorado and Southern Railway (c&s) to bring both beets and coal to their factories. Unfortunately, shipments of both were often inefficient or poorly timed. Sugar beet factories operated during the winter, when coal was in high demand throughout the nation as its most important source of heat. Colorado possessed some of the most productive coal fields in the nation, and c&s provided that coal to the majority of Piedmont families. Though the six factories possessed spur rail lines, they had to lease locomotives from c&s to transport beets. In the winter, transporting coal superseded hauling beets. As a result, beet factories could not achieve efficient shipments. In addition, factories worried about securing the massive quantities of water needed to process beets into sugar.[23] Since all of the region's water had been allocated by 1900, beet factories had to purchase water rights from irrigation companies. In most cases, these were some of the last rights to be satisfied. So, in low water years, they could be left high and dry or dependent on irrigation lawyers to take their case for more water to the courts. Grower unrest, dependency on the rails, and the unpredictable nature of water supply resulted in inefficient management.[24]

Chester Morey recognized this, and he sought to change it. Believing that centralized control of the factories would result in more efficiency and greater bargaining power over workers and resources, he convinced Havemeyer to consolidate all six factories into one company. He succeeded, and on January 13, 1905, the six factories and all of their holdings were consolidated under one management, adopting the name originally used by Charles Boettcher for his Loveland operations: The Great Western Sugar Company. The new corporation was headquartered in New Jersey, initially valued at $20 million, and Havemeyer became its president.[25]

Over the next twenty-five years, Great Western not only consolidated its hold on existing factories but vastly increased the size of its Piedmont empire. By the time of Havemeyer's death in 1907, Great Western built three more factories, in Sterling, Fort Morgan, and Brush, enabling it to control all Piedmont sugar production. Owing to purchases of beet sugar factories in Michigan, Idaho, and Utah during the period from 1901 to 1907, Havemeyer's sugar trust controlled 70 percent of the entire domestic beet sugar industry at his death. By 1930 Great Western had built four additional Piedmont factories. Though smaller independent beet sugar companies operated along the Arkansas River in Southern Colorado, Great Western possessed a virtual monopoly over the Piedmont.[26]

Great Western sought vertical integration as well, achieving limited success. Transportation of beets was of paramount importance. As a raw material, sugar beets are dense and bulky. Growers thus had to consider the time, labor, and expense needed to transport beets to the nearest factory. If the expense did not justify the effort, they would contract to grow only minimal amounts of beets on their farms. Recognizing this, Charles Boettcher, Chester Morey, and the investor cohort who originally built the first factory at Loveland formed the Great Western Railway in 1901. They immediately acquired rights of way and built track to bring growers within easy reach of beet dumps where they could unload their cargo onto waiting train cars for transport to the nearest sugar factory. However, Great Western's dominance was not total. On some routes Great Western failed to acquire rights of way, and so it contracted with C&S to get the beets from farm to factory. In addition, both C&S and Union Pacific lines carried finished sugar to regional and national markets. Despite this, by 1907, when Great Western had finished laying track, most of its Piedmont transportation needs were supplied internally.[27]

Great Western had limited success in controlling the three natural resources it most needed to refine sugar: lime, coal, and water. Turning beets into usable sugar required slaked or burnt limestone to remove impurities from raw beet juice. In 1902 Charles Boettcher and his associates recognized the need to obtain a regular supply of limestone and executed a lease on a quarry in the foothills above Fort Collins. Later known as the Ingleside Quarries, they quickly came under the ownership of Great Western Sugar, which then proceeded to convince C&S to build a spur line to the quarries. The railway completed the line in 1906.[28] Coal was necessary to operate factory machinery and Great Western's locomotives. In this the company saw little benefit in developing its own mines. Obtained through Colorado Fuel and Iron, coal was plentiful and relatively inexpensive.[29]

Water, however, was neither plentiful nor cheap, and Great Western demanded an abundance. In fact, according to its own 1919 figures, during beet slicing operations, which generally ran from October through January, one factory used 6.5 million gallons of water daily, equivalent to supplying the daily needs of a city of 136,000 people.[30] As a newcomer to a region with a well-tapped watershed, Great Western slaked its thirst by building new reservoirs, expanding existing ones, and compromising with Piedmont farmers and municipalities. Because Great Western was always on the prowl for more water, it never came up empty.[31] Thus, regardless of whether the company owned all of the resources necessary for sugar production, its vast capital resources and complete control over refining on the Piedmont made it a virtual monopoly. With the capacity to produce thirty million bags of sugar annually, it was among the largest sugar companies in the world.[32] Great Western had obtained a commanding position over how Piedmont natural resources, water, and labor were utilized.

Beets and Beef: Capitalism in Piedmont Soils

There is a top-down internal logic that seems to follow to this point in Piedmont agriculture during the early twentieth century. The narrative goes like this: The United States sought to assert itself as an imperial power at the end of the nineteenth century through military and economic means. Control of sugar-producing colonies and economic support for the development of a domestic beet sugar industry mapped onto those goals. Entrepreneurs, agricultural scientists, and promoters of all sorts then sought to fill the spaces opened by nationalist goals, tariff protection, and growing sugar demand. Due to its climate, water, soils, and existing agricultural infrastructure, the Colorado Piedmont quickly became a center of production. As independent beet sugar companies gained a measure of success, the Sugar Trust swooped in to take control of its competitors and restore its monopoly over the industry. Piedmont farmers then made a devil's bargain whereby they sold their independent soul to a corporation in exchange for growing a new crop that would make more money. By prostrating themselves to corporate agriculture, they would have to abandon existing agroecological relationships in favor of a farming that sacrificed land health on the altar of profit.[33]

While there is no question that the establishment of a thriving beet sugar industry on the Piedmont was related to skyrocketing consumer demand and the economics of empire and monopoly, on-the-ground relationships between beet sugar industrialists, growers, and the land and water resources of the Piedmont were products of negotiation and compromise. This is in stark contrast to typical turn-of-the-century capitalist narratives, where government-subsidized opportunities for wealth accumulation corroded preexisting social, economic, and environmental relationships. These stories emphasize the value of land and labor as commodities that could be bought and sold, arguing that farmers became one link in a vast commodity chain whereby corporations divided production energies between growers, wage laborers, transportation workers, wholesalers, retailers, research and development, and inputs such as fertilizers, machinery, and chemicals.[34]

Sugar beet agriculture on the Piedmont did not easily yield to that capitalist logic. Cultivating beets required meticulous attention to soil health and crop diversity. Consequently, Great Western Sugar and its growers promoted limited plantings and crop rotation. In addition, most attempts aimed at expansion and efficiency through fertilizers, and chemicals were only marginally successful. Researchers learned quickly that livestock manure was more effective at restoring essential nutrients to the soil than commercial products.[35] Consequently, agronomists and the sugar company encouraged growers to devote time and crop land to animal husbandry. Finally, due to historic patterns of land ownership and the idiosyncrasies of sugar beet biology, Great Western maintained little coercive power prior to harvest. While sugar quickly became the most important commodity on the Piedmont during the early twentieth century, the Piedmont landscape was not so easily commodified.

One reason was that, from an agroecological perspective, Piedmont farmers were relatively successful at the turn of the century. Since the period of accelerated settlement following the Union Colony in 1870, farmers developed diversified farming operations that maintained soil health and included several marketable crops within each rotation. By 1900, virtually all Piedmont farmers grew alfalfa for two to four consecutive years. This restored nitrogen to the soil through its plant roots and, when fed to cattle, materialized as soil-building manure. Cash crops that followed alfalfa varied but included some combination of grains, such as wheat, rye, or barley, and a legume, such as peas or beans. Some Greeley farmers who possessed lighter soils gained a national reputation for cultivating potatoes. While livestock numbers varied widely across the region, the quantity and diversity of feeds grown on the Piedmont were sufficient to support animal husbandry.[36] To supply thirsty crops and animals, farmers tapped the most sophisticated irrigation infrastructure in the American West. Though they clamored for greater water security, their rights authorized them to tap the South Platte River Watershed more thoroughly than any other system of comparable size in the nation. While financial instability on the Piedmont, due to economic depressions during the 1890s, had uprooted some farmers, it did not result from poor land management.[37] Consequently, beet sugar industrialists who sought profit from the economic space afforded by the Dingley Tariff would be hard pressed to reconfigure the existing landscape of agricultural practice.[38]

In fact, that was never their aim. They were aware of agricultural research demonstrating that existing Piedmont farm practices were ideally suited to growing sugar beets. By 1901, when Charles Boettcher supplied the capital to build the Piedmont's first sugar factory, state-sponsored researchers at CAC and multiple divisions of the USDA demonstrated that sugar beets pulled large quantities of essential nitrogen and phosphorous from the soil and required rotation with the same mix of crops Piedmont farmers already cultivated. Alfalfa and legumes, in addition to sufficient quantities of barnyard manure, could resupply nutrients taken by beets. Studies further suggested that growing beets on the same plot of land during more than two or three years out of nine would deplete soil health. Cultivating beets entailed commitment to an entire system of farming. Undermining soil eroded capital investments. Great Western executives and their potential Piedmont growers understood this.[39]

By nature, Great Western Sugar's investment in Piedmont soils necessitated a compact geography. Once refined, the average sugar beet yielded 12 to 15 percent of its total weight in sugar, using technologies that existed in the first two decades of the twentieth century. So, the majority of costs associated with freighting these bulky vegetables would not support sugar production. Further, since sucrose content in beets deteriorates after harvest, time spent in transportation translated to lost profit. The economic fortunes of Great Western and its growers were bound together in Piedmont soils. The company could not simply push its growers to mine the soil and then find other farmers elsewhere to do the same. It needed local farmers—not just some, but the majority—and it needed them to maintain existing practices. Since the company owned little agricultural land, it could not dictate how farmers used theirs. The company's best policy then was to cultivate reciprocal dependency between itself and its growers, and to invest in the health of the landscape on which its bottom line was based. Sugar markets and tariffs made the beet sugar industry, but the local human and soil capital sustained it.

Studies completed during the industry's first two decades on the Piedmont demonstrate the symbiosis of economics and crop rotations. When beet sugar industrialists built a factory in Longmont in 1903, the company's agricultural superintendent, C. S. Faurot, attempted to link crop rotations to profit. He argued that an eight-year rotation including alfalfa, corn, potatoes, peas, wheat, and two years of sugar beets could return a net profit on beets that nearly doubled the money invested. While his claims of profit were inflated, the critical point is that he encouraged farmers to plant beets on only 25 percent of cultivable land, while rotating crops that were already in their arsenal. Studies completed a decade later for three Piedmont beet districts show that growers maintained the crop rotations suggested by Faurot. Even more telling, Piedmont farmer receipts during the period show that between 50 and 90 percent of cash earnings from crops came from selling sugar beets to Great Western Sugar. They continued to replenish soil resources by planting less profitable crops in order to maximize sugar beet gains in only one year out of four. The phenomenal growth of the beet sugar industry is not only a story of transforming a particular crop into a profitable commodity but a story about coordinating energies to sustain a landscape.[40]

Among the organisms that held together Piedmont agriculture and the beet sugar industry, cattle deserve particular attention. By 1900 there was little open range left in the region. Grazing lands on the nonirrigated dryland farms adjacent to the Piedmont were privately owned, fenced, and employed by ranchers for cow/calf operations that raised cattle from birth. Cow/calf operators then sold young cattle to feeders who aimed to profit from the weight increase that confinement and a nutrient-dense diet could provide. Many of these cattle were purchased by Piedmont farmers in the fall, fed over the winter, then sold in Midwestern states such as Iowa and Illinois, where feeders finished them on a diet heavy in corn before selling them to meatpackers in Chicago, Omaha, or Kansas City.[41] For Piedmont farmer/feeders, fattening livestock provided an outlet for alfalfa and other crop residues. Perhaps more importantly, it renewed farmer's fields with the resulting manure. However, from a market standpoint, Piedmont farmer/feeders were at a glaring competitive disadvantage. While the alfalfa that constituted the primary ration for Piedmont cattle could add seven pounds for every one hundred pounds fed, the same volume of corn kernels and silage fed by Midwestern farmers generated double that weight gain.[42] The logical responses were to either add corn to crop rotations or purchase it from Midwestern farmers. Unfortunately, Colorado's growing season was not long enough to cultivate corn in critical quantities and the cost of freighting corn across several states was prohibitively high. Piedmont feeders who shelled out money for Midwestern corn gambled that higher feed costs would be rewarded by a bull market for fattened cattle. However, notoriously volatile cattle markets could result in ruinous losses.[43]

Sugar beet byproducts altered the cattle feeding calculus. Studies completed at CAC in the 1890s and early 1900s, as well as those conducted by Great Western Sugar, demonstrated that feeding large quantities of beet silage and beet pulp could reduce or even eliminate the need to import corn. Beet silage, or tops, referred to the beet crown that workers sliced off at harvest, and the leaves that extended from the crown. For every sixty pounds of beets harvested, growers set aside ten pounds of tops. Left to dry for ten days, they supplied valuable roughage to livestock diets and generated half the weight gain of an equivalent amount of alfalfa. Beet tops could be fed in the fields or loaded into feed bunks, and their availability coincided with the purchase of winter cattle. Tops were concentrated cattle-fattening energy at no additional expense. By 1920 three-quarters of beet farmers fed all of their tops to their own livestock. Beet pulp was even more valuable. It consisted of the fibrous materials that remained after pressing sliced beets in the factory. Studies by CAC and the USDA Bureau of Plant Industry (BPI) concluded that pulp could replace at least half of the corn in cattle rations. Other research suggested that feeders could eliminate corn entirely.[44]

Great Western wasted no time taking advantage of these and other findings, selling pulp to farmers out of its factories; performing animal-fattening experiments at its own experimental farm in Longmont, beginning in 1905; and promoting beet byproducts as feed, while teaching farmers how to use them. Company literature emphasized that beets were "two crops in one," since they provided a lucrative cash crop for the warmer months and livestock feed during the winter months. Constantly aware that beets competed with corn as a feed, company agronomists argued that the locally grown beets could generate 3,500 pounds of sugar and 300 pounds of meat per

Fig. 9. Cattle eating sugar beet pulp. Often mixed with hay, sugar beet pulp provided a cheap and efficient feed that tied Great Western Sugar and its growers together, bolstering livestock feeding practices and supporting healthy soils. Great Western Sugar Company Records, CSU Morgan Library Archives and Special Collections. Used by permission.

acre, productivity unparalleled by the Midwestern Corn Belt. Further, beet pulp and tops were economical and efficient. Into the 1930s, pulp was sold for only fifty cents per ton while tops were essentially free.[45]

It was clear that the research of CAC and Great Western, alongside company promotion, paid off. Through 1919 the local market for beet pulp exceeded supply each year. Statistics for the Longmont District during the winter of 1912–13 alone provide further evidence. According to *Through the Leaves*, a Great Western publication mailed each month to growers, the minority of livestock in the Longmont District that were fed on company property—rather than on grower land—consumed six hundred tons of wet beet pulp daily. The company fattened over nineteen thousand head of cattle on beet byproducts that season for an average of one hundred days each. Those cattle produced an estimated 46.5 tons of manure, capable of fertilizing approximately 1,500 acres and enabling the company to realize a gain in productivity equivalent to $7,700. Beet byproducts and cattle byproducts possessed an elegantly symbiotic relationship.[46]

Jack Maynard played a critical role in clarifying and publicizing that relationship. During the 1920s and early 1930s, Maynard was the associate professor in charge of animal investigations at CAC's experiment station, directing feeding experiments on lambs and cattle that had begun in 1914. Taking his research cue from the beet sugar industry and its growers, Maynard prioritized feeding experiments that employed various iterations of beet byproducts. Though some livestock feeds were grown on CAC's college farm, Maynard contracted with Great Western's factories in Loveland and Fort Collins for pulp and molasses while buying tops and silage from local growers. In addition, Maynard regularly published the results of each year's feeding experiments in *Through the Leaves*. His findings were sweet music to the Piedmont beet sugar industry's ears. Regardless of experiment, Maynard concluded that, when feed costs were taken into account, beet byproducts resulted in the greatest weight gain at the lowest cost. Since Maynard and his students bought and sold the livestock they fed, they were also positioned to provide buyer and consumer feedback. Comparing various feeding regimens, they concluded that rations heaviest in beet pulp yielded the greatest financial returns, with buyers remarking that pulp-fed beef possessed "exceptional flavor and tenderness." Offering further encouragement to Piedmont cattle feeders, Maynard claimed that "cattle fattened in Colorado on sugar-beet by-products, grain, silage, and alfalfa [were] in keen demand at all the principal livestock markets of the nation."[47]

While Great Western Sugar executives were delighted by Maynard's findings and indirect promotion, they recognized that his influence could be more effectively channeled as a company employee. So, in the mid-1930s, Great Western lured Maynard away from CAC, hiring him as their new general livestock consultant. In this role, Maynard scaled back his research efforts, focusing instead on fine-tuning the relationship between Piedmont cattle feeding and sugar beet agriculture. It was with this in mind that Maynard wrote the seminal text on the subject, *Beets and Meat*, in 1945. In this volume, which crystallized research from the previous twenty-five years, Maynard argued for the value of sugar beet agriculture in feeding. He stated that a grower with eighty acres in beets who also fattened one hundred head of cattle for 134 days during the winter season would yield not only cattle for the market, but three hundred tons of manure, enough to fertilize a farmer's crops the following year. Knowing that feeders could choose from a host of products in place of beets to grow their animals, he argued that purchasing large amounts of non-local feeds rarely paid off in the long-term. Maynard exclaimed that feeders in what he called the "beet belt" obtained the greatest success when featuring beet byproducts in combination with locally grown grains and alfalfa.[48] While Jack Maynard's expertise was in feeding livestock, his career served to strengthen the dependencies between Piedmont cattle and the region's soils, crops, growers, college, and sugar company.

The nature of that dependency is rendered more powerful when coupled with the fact that fattening cattle was rarely profitable for Piedmont farmers. For most of the period from 1890 to 1920, American consumption of beef declined and, though market prices for cattle rose and fell, the purchase price for cattle on the open market was not consistently high enough to justify heavy investment. In fact, Jack Maynard's experiments at CAC, despite their endorsement of beet-pulp feeding economics, generally operated at a financial loss. Even with the addition of alfalfa and beet byproducts to their diets, cattle fattened by Piedmont farmers were still finished more often by farmers in the Midwest who sold them to terminal markets such as Chicago, Omaha, and Kansas City.[49] Then, why did Piedmont ranchers and farmers continue to raise stock for the market throughout the period?

The answer is a simple and powerful one. Cattle—and livestock in general—were part of an agroecological economy in which all parts were interdependent. In the absence of significant quantities of synthetic fertilizer, farmers relied intently on barnyard manure, and there was no more prodigious producer than cattle.[50] Given that cattle feeders on the Piedmont provided rations of pulp that could exceed one hundred pounds per day—in addition to their use of beet silage as a roughage—we can deduce that a large portion of this fecal fertilizer proceeding from the backsides of bovines originated from sugar beet cultivation. But, as we have seen, these beets were made possible in part due to a crop rotation that embraced a variety of other plants, the most important being alfalfa, a forage crop that returned nitrogen to the soil and provided an excellent livestock feed. Of course, none of this was possible without irrigation. Here, too, the resource was drawn from the region. By 1900 every last drop of water from the South Platte watershed had been appropriated by users on the Piedmont. Then, from 1900 to 1920, Great Western Sugar and several irrigation companies built or extended a host of water storage projects to add to the region's capacity. And, all of these interlocking pieces were given scientific and promotional boosts by the USDA and the state's agricultural college, which, conveniently enough, stationed itself in the heart of the region and prioritized research that mapped onto the needs of Piedmont farmers. The human and agricultural capital of the Piedmont were largely sustained prior to 1930 on an agriculture that was either derived from or adapted to the region.

Agroecological Economics: Growers and Great Western

For Great Western and its Piedmont growers, land was not just an interconnected web of organisms that collectively held together a profitable industry. It was an essential capital input that increasingly bound the sugar company and its growers together. Since Great Western owned a negligible amount of land and needed to acquire sugar beets locally, it depended on its growers to cultivate a predictable quantity of beets year in and year out. Factories foundered without sufficient supply.[51] In theory, farmers could leverage this knowledge to gain favorable contracts; however, land values and crop choices delimited that position. Productive beet lands with secure water rights increased steadily in value from 1900 to 1930 as water became a more elusive commodity and the Piedmont's reputation as a productive farming region grew. About half of the farmers in the region were tenant farmers subject to high rents, and most landowners possessed mortgages. And, since 30 percent of the cost of raising a beet crop was paid out in rent or mortgage interest, few farmers were positioned to snub the sugar company by refusing to grow beets.[52] No other crops paid out so consistently. Though growers occasionally held back beet acreage to protest unfavorable contracts, they consistently devoted about one-quarter of their cultivable land to sugar beets. When it came to land, neither grower nor sugar company could leverage its position to dominate the resource.

Annual contracts for growing sugar beets reflected this balance of financial risk. Growers assumed all risks prior to factory delivery. Beginning in 1905, Great Western issued annual contracts before the growing season that stated how much the company would pay growers per ton of beets harvested. From 1905 to 1913, that figure was five dollars per ton, with bonuses if the percentage of sucrose content exceeded 15 percent. Typically, a minimum of twelve tons of beets were expected per acre, with that figure generally increasing throughout the period. After 1913 the price per ton rose steadily until the 1920s, when Great Western and its growers agreed annually on contracts with reduced prices per ton in exchange for limited revenue sharing between the company and its growers. All contracts included a stipulation that every harvested beet be delivered to a factory or company beet dump.[53] Though the quantity of beets harvested varied yearly, once annual contracts were signed, Great Western could make broad predictions for how much sugar it might refine in any given year. Further, the company maintained no contractual responsibility for the crop during the growing season. Should a particular grower's crop fail, Great Western's factories would lose beets to process; but it bore no burden to insure its growers against losses to either crops or land, or to pay hand laborers their wages. In such scenarios, growers and laborers experienced losses far greater than those sustained by the company.

Great Western still assumed risks related to processing beets, the sugar market, and its capital investments in Piedmont agriculture.

Once growers signed contracts to plant a given number of acres, Great Western could plan for its refining operations and gained information that would help it forecast distribution of refined sugar. By contrast, the company possessed no security against the volatility of the market. It negotiated annual contracts with growers, usually in March, for a crop that would not be ready for sale as refined sugar for at least ten months. Though the company could occasionally hold back some sugar when prices flagged, this was not typically an option due to its obligation to growers and investors. Should the company withhold its product within its regional market on the plains and in the Midwest, company executives feared that sugar from elsewhere might move in permanently to fill the vacuum.[54]

Great Western's network of sugar beet research and dissemination represented another layer through which it divided the energies of making sugar among specialists. Beginning in 1905, the company established research stations at Longmont and Fort Collins, where the company's scientists investigated improved methods in beet breeding, farm management, pest control, irrigation, sugar beet pathology, and beet growing economics.[55] To disseminate their findings, the company employed a host of trained agricultural workers called fieldmen who acted as company mouthpieces and general consultants to growers. As of 1919, Great Western employed seventy such fieldmen on the Piedmont.[56]

Lowell Giaque and Norman Vlass typified Great Western fieldmen. The two grew up on beet-growing Piedmont farms during the 1920s and 1930s, went to college at CAC, and worked as fieldmen for Great Western during the 1940s and 1950s. Giaque recalled that, by the time he went to college in the early 1940s, CAC provided a pipeline to employment with the sugar company. While at school, he and many of his classmates worked part-time at night during the winter refining season and were paid "good wages." Immediately after graduation, he went to work full-time for Great Western in Sterling. Norman Vlass was recruited by Great Western while still a student at CAC. Vlass stated that, after graduation, working as a fieldman was a natural extension of an experience growing up in a family that cultivated sugar beets. In the spring, Vlass walked the fields with farmers, assisted with proper measurement of acreage, and advised them on the season's labor, irrigation, and cultivation. During the season, he informed them about pests and diseases that threatened the crop and recommended the use of specific pesticides and fertilizers. At harvest, fieldmen coordinated the delivery of beets to local beet dumps. Giaque and Vlass also collaborated with county agricultural extension agents and conducted farm demonstrations. Despite their visibility, Vlass pointed out that fieldmen were successful when advising and assisting, but never when directing. When identifying lines between growers and fieldmen, Vlass laughingly recalled that German Russian farmers informed him of his transgressions by switching from English to German, thereby cutting him out of the conversation. Vlass and Giaque's work as fieldmen underscores the mutual dependency that defined Great Western's relationship with its growers.[57]

During the early twentieth century, the beet sugar industry planted all of the elements of modern industrial capitalism onto the Piedmont. Beet sugar industrialists and regional farmers maximized their capital inputs into producing a commodity—sugar—for the market. With the goal of wealth accumulation, Great Western invested profits into physical infrastructure, research and development, and company propaganda. These vaulted the company into first place among domestic sugar producers. With the economic bottom line in mind, Great Western made efforts to offload financial risks onto growers and laborers. It developed a sophisticated division of labor to include growers, an exploited set of migrant field laborers, factory workers, researchers, and company executives. In addition, the company actively subsidized its success through the expertise of state-sponsored scientists from the USDA and CAC. And, supporting the weight of the Piedmont's sugar capitalism were federal tariffs that protected Great Western's economic viability.[58]

Yet, the trappings of modern capitalism were not capable of dulling the particularities of place on the Piedmont. A central industrial ideal in the early twentieth century was to pull material resources from multiple locations of extraction so that they could neutralize the value of any single site of production. If sugar beets grown in one locale could be efficiently transported and refined elsewhere, then Great Western could leverage that advantage in its economic dealings with Piedmont growers. The company could then acquire beets from several regions and minimize the need for regenerative agriculture in any single one. This would, in turn, neutralize the power of farmer land tenure since each plot of land could be treated as an interchangeable part, expendable and replaceable by the same part produced elsewhere.

This attempt to reduce human and ecological relationships to faceless capital inputs foundered on the Piedmont landscape. The most complex interactions taking place in the beet sugar industry occurred not at the level of factories or finances. Rather they were contained within the soil. The work of farmers and wage laborers to cycle critical plant nutrients back into the earth, made possible by well-ordered crop rotations and copious quantities of livestock manure, could not be reduced to an expendable factor of production. Sugar beets could not simply be grown in several regions and then transported to the company's factory of choice. Efficient cultivation and refining were bound by prescribed geographies and local soils. Further, land was not a resource the company could manipulate since historical patterns of land tenure and water distribution on the Piedmont—and through most of the sugar beet growing regions of the American West—gave farmers autonomy over what they planted. While the beet sugar industry on the Piedmont, and throughout the West, refined a generic commodity subject to complex and shifting markets, its principal capital was fixed in place, within local human and biological relationships.

The example of the beet sugar industry on the Piedmont from 1900 to 1920 also suggests the constraints under which capitalist structures can operate to support healthy agroecologies. By 1910 all of the trappings of modern industrial capitalism existed in the beet sugar industry: specialization, division of labor, economies of scale, expertise, wage labor, a rationalization of the labor process, lobbying of political officials, wealth accumulation, and investment of profits into expanded production.[59] Despite all of this, Great Western Sugar was constrained by capital inputs that were available locally. The company could only expand its grower base within a narrow geography due to the high cost of transporting beets, the loss of sugar content in transit, and the fact that the water necessary to grow this vegetable was largely tapped out. Without significant investments in water infrastructure, this would always be the case. The same limits applied to growers as they had no economically feasible choice but to sell their beets to Great Western. Further, since they could not grow corn efficiently, or transport it cheaply from the Midwest, Piedmont farmers were stuck with beet pulp. This cheap but second-rate feed supported limited meat and manure production but could never be a moneymaker under existing constraints. Piedmont growers were bound together with their corporate sponsor by a shared investment in the agricultural health of the region's land. Capitalism was restrained by what was possible on the Piedmont landscape even as it functioned to sustain its existing agroecology.

One glaring exception existed to this dependence on Piedmont lands and resources. Cultivating sugar beets was an intensive operation that required human muscle, which growers and Great Western extracted from the bodies of migrant laborers. Though the lives of these workers appeared to be transient, their energies were baked into Piedmont agroecology. During the first half of the twentieth century, the Piedmont beet sugar industry conserved local resources while exploiting imported laborers. For the sugar company and its growers, it seemed impossible to do the former without the latter.

3 Beet Biology and the Nature of Labor

> Every day, sunup to sundown . . . work all day, eat what you
> can, crash for the night. Do it again the next day.
>
> —ELVA TREVIÑO HART

Managing soils and managing crops are related yet distinct operations that employ different kinds of energies. So it went with sugar beet farming. As Piedmont farmers and Great Western Sugar observed, the right combination of rotting manure, beet byproducts, and restorative plants, if given time, replenished the organic life and nutritive content that upheld their industry. A crop species, however, does not operate on that logic. The singular and independent DNA that is locked into each seed exists apart from the soil in which it is planted. Once germinated, that DNA is expressed in a world of subterranean earth, sunlight, and water. For sugar beets, the genetic material locked within these sucrose-rich vegetables contained the seeds of extra-local labor exploitation even as they helped to root Piedmont agroecology in place.[1]

Prior to the 1940s, the term "beet seed" was misleading. By nature, sugar beet seeds, like all beet seeds, exist in multigerm seedballs that usually contain from two to five embryos. When each embryo germinates, it is part of a root ball whose shoots break the surface as several tightly compacted and sometimes intertwined plants, each one competing for water, nutrients, sun, and growing space. The first step in transforming this vulnerable cluster of seedlings into bulbous,

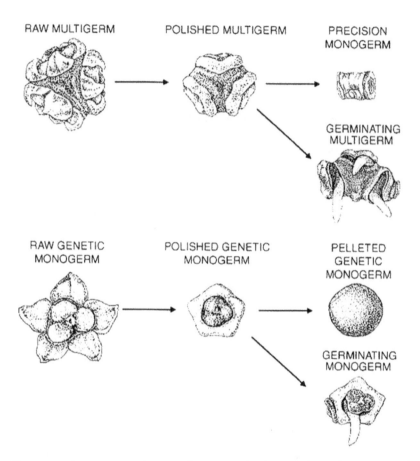

RAW MULTIGERM

POLISHED MULTIGERM

PRECISION
MONOGERM

GERMINATING
MULTIGERM

RAW GENETIC
MONOGERM

POLISHED GENETIC
MONOGERM

PELLETED
GENETIC
MONOGERM

GERMINATING
MONOGERM

Fig. 10. Multigerm sugar beet seed. Since multiple sugar beet plants emerged from a single, multigerm seed, they required extensive thinning to leave healthy, single plants. This was the biological basis for the size and persistence of the contract labor system in the beet sugar industry. Graphic from Biancardi, Panella, and Lewellen, *Beta Maritima: The Origin of Beets*.

sugar-rich vegetables involved painstakingly removing weaker plants, leaving space for the healthiest ones to thrive. This need for thinning on a massive scale explains why Great Western Sugar imported tens of thousands of contract laborers annually to cultivate the beets that formed the foundation of its sugar empire.[2] Beginning in 1901, when Great Western built its first sugar factory in Loveland, the company and its growers sought out entire families of unskilled laborers who existed at the fringes of society, willing to do exhausting labor for long hours and low pay. Using a bevy of labor agents, Great Western employed entire families of German immigrants from Russia, small numbers of Japanese Americans and, after World War I, predominately Hispanos and Mexican nationals.[3]

Contract laborers and their families were essential to beet sugar capitalism and to Piedmont agroecology. Well into the 1930s, labor represented more than half of all costs associated with cultivating sugar beets, and the hours expended by fieldworkers vastly exceeded the combined hours contributed by growers and Great Western Sugar's employees.[4] The toil and sweat of men, women, and children were embedded in the region's soils, plants, and water. Beet agroecology was inseparable from the work of cultivation. The lifecycle of sugar beets and the particularities of local soils structured human labors. Visual cues and muscle movements synchronized with the rhythms of sugar beet lifecycles. The same water that fell from the sky and irrigated fields attracted mosquitoes, seeped into homes, and harbored disease. Children as young as six were notably impacted since their young bodies engaged in all of the same cultivation processes as their parents. Repetitive and demanding movements, some of which utilized hooks and knives, retarded musculoskeletal development and resulted in both chronic and acute injuries. These demands also stunted intellectual growth since thinning and harvesting fatigued children and prevented school attendance, thus supporting the derogatory stereotypes Anglos commonly held about migrant laborers.[5] Piedmont agroecology also penetrated the grower-owned dwellings occupied by families of workers in the form of dust, insects, rodents, and drinking water. Moreover, all laborers bore the marks of long days of thinning, pulling, and slicing beets, on their bodies and within their muscles and joints. In this way, the impacts of sugar beet agroecology were not bounded by the time and space of a particular season.[6]

As laborers attested, cultivating sugar beets was fraught with unpredictability and environmental risks that existed in tension with capitalist prerogatives. While contract laborers and their families adapted their efforts to accommodate nature's whims, Great Western and its growers rationalized nature by shifting environmental risk onto its laborers. By planting early, delaying harvest, and taking chances with limited irrigation, growers and the sugar company hoped to sweeten profits. These tactics, however, invited volatility. So, to minimize consequences, they developed contracts that shifted the burden of these risks onto the shoulders of laborers and their families who not only endured miserable working and living conditions but loss of income as well. In agro-capitalist relations, all parties seek a farm ecology that is orderly and predictable—goals that can never be entirely realized. Great Western and its growers structured labor contracts to offload some of nature's irrationality onto its workers.[7]

Finding Labor: German Russians

Sugar beet agriculture on the Piedmont offers a clear example of the choices and contingencies involved in attracting a labor force. The Chemistry Division in the U.S. Department of Agriculture (USDA), under the leadership of Harvey Wiley, performed experiments on the suitability of sugar beets for American soils since 1880. While promoting the vegetable, the USDA acknowledged that beets required a large labor force.[8] During the 1890s, as experiment station researchers at Colorado Agricultural College (CAC) performed a host of experiments on the suitability of sugar beets for the Piedmont, they admitted that the crop required labor unavailable to small farmers. They assured them that beet sugar industrialists found laborers wherever the vegetables were grown in the past and that they could be found in Colorado from among a local pool.[9] This was willful deception. The overwhelming majority of Piedmont residents either owned their own farms, rented land, or worked in local businesses, many of which depended on the region's farm economy. Moreover, the heavy physical demands and temporal nature of hand labor in sugar beet agriculture precluded finding a critical mass of locals willing to engage in it. Beet boosters and farmers in Piedmont communities were aware of the need for a labor force that did not exist even as they promoted the industry. They were conscious that cheap labor would have to be imported from society's margins, owing to its physical demands and seasonal nature. Thus, even as Great Western Sugar and its growers embraced a symbiosis between local, sustainable farm practices and cultivating beets, they also developed an economic model rooted in labor exploitation.[10]

Even as high tariffs protected the fledgling beet sugar industry from foreign competition, growing nationalism and international upheavals brought immigrants into the labor-intensive agricultural industries of the American West at the turn of the twentieth century.[11] Between 1901 and 1920, Great Western and its growers took advantage of global tensions by drawing German Russians into the beet fields. By the 1880s, large populations of Rhineland Germans had been living in the treeless Russian steppes near the Volga River for well over a century. Originally enticed by Catherine the Great in the 1760s with offers of free land, political autonomy, and exemption from military service, they found themselves squeezed in the vice of Czar Alexander III's Russification policies. Stripped of their former benefits and required to assimilate, many opted to emigrate. German Russians were lured to the Great Plains by promises of cheap land, a terrain and agriculture analogous to the Russian Steppe, the ability to immigrate as communities, and the promotional efforts of American railroad agents operating in the Volga. Families of German Russians began to arrive in plains states such as Nebraska and South Dakota in the 1880s.[12] Since land was not as cheap and available as they were led to believe, many found jobs on the railroads or moved into cities such as Lincoln and Denver. As their communities stabilized, they paid for the transport of relatives and friends, with sizable migrations occurring from 1890 to 1892, and a much larger influx from 1898 to 1913.[13]

The background and timing of German Russian immigrants enabled them to adapt to the needs of the new beet sugar industry. In 1889 Henry Oxnard opened a refinery in Grand Island, Nebraska, and local farmers pledged to plant several thousand acres in beets. German Russians provided a steady pool of labor for the industry at the same time that employment on the rails slackened. Many hoped that hard work in the beet fields would yield enough money to buy the sort of land that had attracted them to the plains initially. When Great Western opened its first factory in Loveland in 1901, it was happy to transport many of these new immigrants to the region's sugar beet fields.[14]

German Russians provided an ideal source of labor for a fledgling industry that sought cheap, exploitable workers. As newcomers from Eastern Europe, they lived on the margins of an American society that viewed them with suspicion.[15] In addition, they possessed some familiarity with sugar beet agriculture, having farmed root crops such as potatoes and beets in Russia and on the plains of Nebraska. Further, since German Russians were motivated by future landownership, they willingly employed their entire families in the sugar beet fields. Great Western's labor contract reflected this reality since it did not clarify who would complete the work. Contracts stated the number of acres that laborers would thin, weed, and harvest, receiving compensation following the season's work. Here again, German Russians personified industry needs since families tended to be large, averaging two adults and five children, enabling them to contract for at least fifty acres, earning an average of $500 for the season. With frugal living and some off-season employment—often hard to come by—families working together in the fields could realistically become landowners within ten years. Thus, it appeared on the surface that the Piedmont's beet sugar industry sustained fair wages for its farm laborers. In truth, those earnings were borne on the backs of entire families.[16]

Contract labor and field conditions followed workers into their homes. While Great Western recruited and often transported workers to the Piedmont, they left housing arrangements to growers. In 1901 Great Western's first year on the Piedmont, fifty-four German Russian families contracted to complete some of the necessary hand labor. Eight years later, eleven thousand laborers—mostly German Russians—worked in Piedmont beet fields.[17] Since hand labor required over ten hours each day in the field, and there was little or no housing available nearby, most beet workers lived on-site. To save money, growers developed all sorts of makeshift housing, from providing spaces in their barns, to tents. But, within a few years, most grow-

Fig. 11. A German Russian family of nine working as contract laborers in the sugar beet fields near Greeley in 1915. Photo by Lewis Hine, Records of the National Child Labor Committee, Library of Congress, Prints and Photographs Division.

Fig. 12. Typical beet shack near Sterling in 1915. Photo by Lewis Hine, Records of the National Child Labor Committee, Library of Congress, Prints and Photographs Division.

ers erected dwellings on their properties, derogatorily named "beet shacks." A typical shack was 14x20 square feet, with 336 square feet of living space. It possessed two rooms. One functioned as kitchen, dining room, and living area. The other was a bedroom intended for the entire family. Growers enclosed shacks on all sides, cut two windows, laid rudimentary wood plank floors, and constructed tar paper roofs.[18] Housing clarified the place of hand laborers. Tight spaces, cheap construction, and immediate proximity to beet fields left no doubt that their economic value resided solely in their labor. It also suggested that contract labor and Piedmont agroecology were inseparable.

Living and working arrangements varied from the end of harvest until mid-spring, when laborers returned to the fields. Since most German Russians aimed to purchase Piedmont lands, they generally tried to remain in the region. Some were fortunate enough to find additional work for Great Western—unloading beets at factories—though their immigrant status defined them as not "American" enough to rate employment inside factory walls. Still others found work on Great Western's network of rails. Those who resided in Denver sought work at one of the city's smelters or sorting potatoes and beans in an agricultural processing facility. Women occasionally found work as domestics. While the nature of housing varied, German Russians invariably lived in ethnically homogenous neighborhoods segregated from the rest of the population. Thus, regardless of whether German Russians found work during the off-season, Great Western's labor recruiters could locate their labor pool in the spring.[19]

Though German Russians performed the majority of hand labor in the sugar beet fields for most of the first two decades of the twentieth century, their family labor, along with international events, increased wages over time, and relatively affordable land enabled many of them to become growers, depleting the labor force even as the demands of a growing beet sugar industry expanded. A common phrase German Russians carried from Europe was "Arbeit komm hier, ich fress, dich auf!" ("Work, come here, I will devour you!") That attitude motivated long hours of labor and the employment of entire families in the beet fields. Having arrived in America seeking agricultural land, many utilized their wages to move quickly from laborer to tenant to owner.[20] That process was at its most pronounced during and immediately following World War I. The eruption of the Great War in 1914 immediately cut off immigration from Europe, resulting in a labor shortage. The war also decimated Europe's beet sugar industry, creating a global shortage even as sugar was in high demand by troops in the conflict.[21] With sugar and labor in short supply, fieldworkers commanded higher prices for each cultivated acre, a trend that continued for two years following the war.[22] This propelled even more German Russians into land ownership. As its European pool of immigrants waned, Great Western sought labor closer to home.

"A Permanent Underclass of Labor"

Great Western recruited Hispanos and Mexican nationals to fill the labor vacuum. This was not without precedent on the Piedmont, but the scope and nature of the recruitment would have profound consequences for the region's labor regime and the construction of race. In 1903 Great Western recruited 275 Hispanos from the American Southwest to increase the labor supply at the industry's Piedmont genesis.[23] For the next decade, small groups of these internal migrants worked in the beet fields, with most of them returning to their communities in southern Colorado and northern New Mexico during the winter season. When World War I constricted the European labor pool, Great Western actively targeted Hispanos to fill shortages. Unable to attract significant workers within U.S. borders, Great Western imported Mexican nationals. Initially the company recruited single men who contracted to either thin, cultivate, or harvest beets. However, by the end of the war, when it became clear that these new laborers could fill the void left by German Russians, Great Western recruited entire families. By 1920 an even number of Hispanos and Mexican nationals worked in the beet fields, providing 40 percent of hand laborers and 90 percent of all migrant workers. During that year, Great Western employed twenty full-time recruiters, spending nearly $200,000 to transport workers to the beet fields. By the mid-

1920s, Mexicans comprised over two-thirds of the hand labor force, and Great Western employed thirty-five labor recruiters who operated in the Southwest, Texas, and northern Mexico.[24]

Immigration status played a significant role in the transition, offering insight into how racial attitudes permeated labor. During the Progressive Era, attitudes about the influx of Eastern Europeans centered on perceptions of assimilability into American society and overblown concerns that some immigrants endangered democratic institutions or would become wards of the state.[25] These attitudes informed the Immigration Act of 1917, which contained an English literacy requirement and a doubled head tax on immigrants. Piedmont residents typically viewed German Russians with suspicion since most still spoke German as their primary language in the 1910s and many refused to cooperate with the school system. This fueled fears of German Russian disloyalty to the United States during World War I. However, the new immigration law and wartime discrimination pressured most German Russians into learning English.[26] Moreover, the fact that so many of them were able to purchase land convinced detractors of their self-sufficiency and fitness for democracy. Their second-class status had revolved around their marginality in the labor regime. They became Americans in the transition from hand laborer to grower.

The Johnson-Reed Act of 1924 further highlighted the transition within the German Russian community from contract laborer to grower. This statute placed quotas on European nations based on populations of immigrants and their ancestors living in the United States in 1890. This meant that nations such as England, France, and Ireland that had historically contributed significant numbers of immigrants to the United States by 1890 were allotted large quotas. Since German Russians emigrated from Russia, they were grouped with the quota on Russia, a nation that had contributed few immigrants by 1890. The Johnson-Reed Act virtually eliminated German Russian immigration, thinning their laboring ranks as they transitioned into growers.[27]

The manner in which immigration laws addressed migrants from Mexico demonstrates how Great Western and its growers made racial and labor distinctions between German Russians and Mexicans. Great Western was one among many companies in the West seeking cheap labor—primarily in agriculture, mining, and on the railroads. As the labor market tightened during the war, these companies successfully lobbied Congress to eliminate literacy tests and head taxes for migrants from Mexico in the 1917 Immigration Act. Though these exceptions were intended to last only a few seasons and apply only to temporary workers, rules were rarely enforced and many Mexican nationals deliberately chose not to return to Mexico at the end of their contract period.[28] By the mid-1920s, one-third of all Mexicans in the United States worked in agriculture, comprising three-quarters of all laborers in southwestern fruit orchards and vegetable fields.[29] Immigration law and the exceptions crafted by growers and Congress both responded to and incentivized demands for an abundant and exploitable Mexican labor force.

Whereas Great Western and its growers made distinctions for German Russian laborers as to whether they were capable of assimilating, the sugar company recruited Mexicans under the assumption that they would never assimilate. A 1920 U.S. Department of Labor study concluded that Mexicans "are not permanent, do not acquire land . . . but remain nomadic and outside of American civilization."[30] In other words, unlike German Russians, Mexicans could always be relied on to do the backbreaking, low-wage work in the fields, and then disappear when their services were no longer required. In 1920, as Congress debated immigration restrictions, representatives of the beet sugar industry weighed in. They argued that restricting Mexican immigration would be the death of their industry, striking a significant blow to American agriculture since white Americans resisted the physical labor assigned to Mexicans. Highlighting the perceived distinction between Mexicans and whites, a representative of the beet sugar industry concluded, "You have got to give us a class of labor that will do this back-breaking work and we have the brains and the skill to supervise and handle the business part of it."[31]

Great Western and its growers affirmed that Mexicans fulfilled their desire for a permanent underclass of laborers. As the chief labor recruiter for Great Western, C. V. Maddux was often called upon to defend this philosophy. Appearing before the House Committee on Immigration and Naturalization in 1928, he stated that he was not worried that Mexicans would ever supplant American farmers. In fact, he implied that this was one of the reasons why Great Western actively recruited them. Knowing that most of the beet lands of the Piedmont were already in production and that there was little room for more growers, Maddux stated, "We no longer want settlers to occupy vacant land . . . what we want is workers to work for the set-tler who came before."[32] Great Western and its growers substantiated their arguments that Mexicans would remain laborers by distilling observations into stereotypes that pervaded the beet sugar industry. Viewing that their families were smaller than those of the German Russians before them and that Mexican mothers and children worked fewer hours in the beet fields, growers argued that Mexicans did not want to move ahead and were unwilling to help themselves in times of trouble. Observing that they had not moved into land ownership, others stated that Mexicans were not thrifty and would not "save for a rainy day." In his 1929 study of sugar beet labor on the Piedmont, noted University of California agricultural economist Paul Taylor quoted a "college trained farmer" who summarized the prevailing relationship between Great Western, its growers, and Mexicans: "Beets are largely responsible for the development of this country, and beets require Mexicans."[33]

Great Western and its growers employed selective memory in clas-sifying the Mexican labor force. From its inception, the beet sugar industry, growers, and boosters on the Piedmont argued that a local labor force could be found to do the work of cultivation, a claim based on a willful ignorance of the lack of available workers on the Piedmont. By the 1920s, Great Western not only actively recruited distant laborers but also employed a lobby in Congress to maintain its cheap labor supply. Once in place, the company and its growers labored to keep workers on the margins. This presented a particular irony regarding the new Mexican labor force. The evidence used to argue that Mexicans were unwilling to help themselves or move ahead in the world was taken from their tendency to work fewer hours and acres than their German Russian predecessors. By contrast, those same German Russians were previously disparaged for assuming large workloads and placing entire families into the fields, enabling them to buy land and become growers. Unfortunately, no matter how diligently the new labor force worked, opportunities to become growers were sorely constricted.

While it was true that Mexican families were smaller on average than their German Russian predecessors, and that Mexican women and children worked fewer hours—though family labor was still the norm—several factors beyond their control kept them from becoming growers.[34] One was the simple fact that the irrigated lands of the Piedmont were among the most expensive agricultural lands in the interior West, and their value escalated quickly in the post-war period. Writing in 1919, Great Western executive J. F. Jarrell stated that demand for beet lands was so high that some had doubled in value within two years.[35] In addition, Mexicans who arrived on the Piedmont were often saddled with debt from the start. Though Great Western paid for their transport, laborers often appeared a full month before thinning season started. To enable the new laborers to establish themselves, Great Western, growers, and retailers advanced money to beet workers until the end of the season when they were finally paid in full. Laborers then paid off their debts and routinely left the fields at the end of the season with little money in their pockets, returning the following year in need of further advances. Sometimes called the *padrone* system, this cycle of debt placed insurmountable walls in front of laborers who hoped to advance their status.[36]

Despite poor living and working conditions, it is impossible to ignore that Hispanos and Mexican nationals still chose to come in droves to work in Piedmont sugar beet fields. Why? Historian Sarah Deutsch showed that many Hispanos selected work in the beet fields because of the social disruptions caused by World War I. Some Hispano males fought in the war, while others became migrant laborers, seeking higher wages in the mines and agricultural fields of the Southwest. The war and the heightened visibility of Hispanos energized discrimination against them, limiting job opportunities outside of agriculture. Labor recruiters from Great Western offered plentiful work after the war at wages that were higher than they could make in their towns in southern Colorado and northern New Mexico. And, the contract labor system, exploitative as it was, allowed entire families to work in the beet fields.[37] Among Mexican nationals, many chose to migrate in the wake of the tumultuous Mexican Revolution. Though that conflict purportedly aided many on the lowest rungs of Mexican society, it nonetheless created hardships for Mexicans from all social backgrounds. Further, Great Western's labor recruiters working in Mexico promised more benefits than work in the beet fields could deliver. Once communities of Hispanos and Mexican nationals formed within Piedmont towns, newer migrants could find safety and solidarity in existing enclaves. Finally, well-organized charities and state and federal agencies, provided meager economic safety nets during hard times. While Mexican nationals and Hispanos were aware of their second-class status on the Piedmont, work in the sugar beet fields offered tradeoffs that many willingly accepted.[38]

Entangling Beets and Bodies

The contract labor system employed by Great Western and its growers in Piedmont sugar beet fields was far more than an exchange of productive labor for monetary compensation. The agroecology of an entire region was baked into the individual bodies, family lives, dwellings, and societal relations of its principal laborers. Beet labor was embedded in the weathered skin and aching joints of those who stooped, crawled, sliced, and selected for up to fifteen hours daily under the intense glare of the Piedmont sun. It was legible in the public schools where the rhythms and idiosyncrasies of cultivation and harvest circumscribed the educational opportunities of migrant laborers and illuminated the tangled relationship between race and child labor. Sugar beets also permeated workers' homes. The hastily constructed beet shacks, situated adjacent to fields, provided little respite from the physical world of wind, rain, dust, insects, and disease. In visceral ways, the materiality of sugar beet labor presided over the indoor spaces of families. Contract laborers in Piedmont beet fields were both shaped by and fully integrated into the region's agroecology.

The labor of sugar beet agriculture—on the Piedmont, the coastal plains of Southern California, and the Upper Midwest—was rooted in the biology of the plant itself.[39] From the moment that beets germinated and emerged from the earth, the sugar clock ticked. Since seed germs could contain as many as five seeds, plants burst the soil in a tight asymmetrical arrangement, each one competing for soil, water, air, and sunlight. Left alone, they were incapable of yielding the impressive tubers that would be harvested and processed five months hence. To thrive, all but the healthiest plants required elimination. Further, while growers employed horse-drawn seed drills to plant beets and cultivators to cut down weeds between rows, thinning within each row defied mechanization since existing technology was predicated on uniform spacing. So, while agricultural operations rapidly replaced human muscle with machines and fossil fuels during the early twentieth century, beet growers depended on human labor to perform most tasks.[40]

This began with blocking and thinning, starting in late May. Blocking involved cutting out large sections of beet plants entirely, while precision thinning required the removal of the weaker seedlings in the clumps that remained. Those tasks were performed simultaneously as laborers passed up and down rows. Depending on weather and soil conditions, a typical contract worker; his wife; and three children, working alongside their parents, could block and thin as many as thirty-five acres in six weeks.[41] In a process that some laborers described as stoop-walking, they moved rapidly between seedlings, chopping out entire plant sections with a short handled hoe and thinning the weakest remaining plants by hand, all the while standing, stooping, and kneeling to obtain the most advantageous position for the task. Rows were typically one-quarter to one-half mile long, and each acre totaled approximately five miles. With clear admiration for the skill, rhythm, and fortitude required to thin beets, Elva Treviño Hart described her father's labor this way: "He worked two rows at a time . . . slicing three times to the left, then three times to the right, like a dance." The goal was to leave the strongest plants standing, spaced ten to fourteen inches apart, while identifying and eliminating weeds.[42] When completed with alacrity and precision, no process proved more essential to an abundant harvest of bulbous, sugar-rich beets than blocking and thinning.

Once thinned, could the remainder of beet cultivation be mechanized? After all, with plants evenly spaced, it made sense that an accurately calibrated machine cultivator could efficiently remove weeds emerging between plants. While growers and research institutions such as CAC solicited and developed machinery that could cut out weeds between beet plants, hand-weeding operations remained the norm until after World War II on the Piedmont. Why? Machines could only be calibrated to the degree that beet spacing was uniform; however, the most vigorous plants rarely germinated at predictable intervals. Only single seeds, planted at uniform distances, made that possible. Then there was the issue of cheap labor. Great Western and its growers were able to attract contract laborers and their families in such large numbers precisely because they could guarantee work for an entire season. Eliminating labor phases piecemeal could unravel the entire system. Finally, there was cost. Growers were reticent to invest in machinery, such as tractors and attachments, that could only be used in selected operations. To gain maximum utility on the relatively small acreage employed by most beet growers, a tractor that could be fitted with attachments for multiple phases of beet cultivation was needed. Only individual beet seeds could make that possible. Prior to 1940, this was not on the horizon.[43]

Workers commenced hoeing weeds immediately following blocking and thinning. Even an immaculately thinned field of beets required at least one round of hoeing since many weeds had not yet germinated during thinning operations. Performed under the glaring July and August heat, laborers passed through fields, identifying and eliminating all plants that were not beets. German Russians stood and used long-handled hoes for weeding while Mexicans typically wielded short-handled versions on their knees. The process, often called second-hoeing, might require only one pass through the rows, but often necessitated more, depending on the skill of the laborers, the vagaries or weather, and how well the grower had managed his land throughout the year.[44]

In August, once beet plants attained the size and vigor needed to thrive, weeding operations ceased, idling laborers until harvest. In some beet-growing regions, such as California and Minnesota, migrant laborers could obtain interim work in one of several crop harvests.[45] On the Piedmont, fewer opportunities existed, largely due to land ownership patterns and sugar beet agroecology. Owing to original settlement patterns dating back to the Union Colony, Piedmont farmers owned or leased, on average, fewer than one-hundred acres each. Outside of the intensive cultivation required for sugar beets, most could manage their crop rotations and modest numbers of livestock without hiring additional labor. In addition, there were no significant opportunities in the regions adjacent to the Piedmont. To the west lay the Rocky Mountains. On the plains to the east, where grains predominated, mechanized cultivation and harvesting had replaced human muscle by the early twentieth century. While some Piedmont beet laborers could find temporary work on the railroads or piece work on farms, their sheer numbers combined with limited opportunity kept most of them in place until the harvest. The particularities of sugar beet biology brought workers to the Piedmont in large numbers and held them there between cultivation and harvest operations.

The need for armies of hand laborers during the sugar beet harvest, which commenced at the start of October, was also heavily determined by beet biology. Since sugar beets possessed a taproot that could extend several feet into the ground, removal was no small task. During the first two decades of Great Western's operations, workers extracted beets entirely by hand. Over the course of the 1920s and 1930s, the beet sugar industry, land grant colleges, and the USDA collaborated to develop machines that could first loosen them from the soil and eventually extract them entirely. During that period, they also developed mechanical loaders that moved beets from field to truck. Despite this, the most time- and labor-intensive portion of the harvest, called topping, remained in the skilled hands of laborers.[46] Workers wielded a knife with a four-to five-inch wooden handle and a sharp, rounded blade that extended up to fourteen inches from the handle. A small hook protruded from the top of the blade. As beets were pulled, workers stooped at the waist and, in one motion, thrust their hooks into each beet, one at a time, hoisted them, and removed them from the hooks. Laborers then topped beets by hand. Bracing them against their knees, they forcefully sliced through the crowns of each vegetable with their blades, tossing the beets and tops into separate piles, where they would alternately be processed into sugar or provide feed for livestock.[47]

From April through November, two out of every three hours required to move sugar beets from seed to harvest were performed by hand laborers in the fields.[48] The time and exertion they expended inscribed beet agroecology on their minds, senses, and muscle memories. Take their experience of water and soil. During the thinning season, water falling from the sky above and percolating from the earth below dictated labor possibilities. Families often worked in thick clay mud that attached itself to shoe soles and hoe blades, adding to the physical toll of walking, stooping, and kneeling. It was common to emerge from a long day caked in mud. These physical exertions were prolonged by Piedmont weather patterns that delivered the season's highest rain totals in May. The practical economics of mud were left to the male head of the household who made daily decisions about whether field labor was possible, while the extra burden of restoring clothing's field utility fell to adult women. Moisture also lured mosquitos drawn to wet fields and human sweat, prompting workers to cover their faces and necks with towels and handkerchiefs. The dry weather of July and August thinned mosquito populations, except during several irrigation periods when their numbers swelled. While dry conditions made work more palatable, it generated its own problems. During weeding operations, Saul Sanchez complained that "the soil was so hard and the weeds were so thick that the hoe bounced right back when it hit the dirt," adding to the bodily exertion and physical toll of the labor. Freezing and thawing water required endurance and care as well. Frosts frequented the Piedmont beginning in late September, coinciding with beet harvests. During early morning hours, topping demanded extra care, as icy beets conspired with sharp knives, causing injuries to knees and limbs. As the weather warmed through the day, ice turned to water, soaking workers from the knees down. One witness observed that women were "wet up to their waists and [had] ice in their laps and on their underwear."[49] Sugar beet agroecology and the field labor on which it depended were deeply embedded in one another. The visceral connection between human bodies and earthy soils bore out that relationship.

The demand for families of contract laborers grew astride the beet sugar industry, catching the attention of local activists, national progressive organizations, academics, and agencies of the federal government. Their ensuing investigations and publications, beginning in 1916, were then picked up by various media outlets who processed findings for the reading public.[50] No issue elicited more attention and public outcry than child labor, as multiple studies documented its pervasiveness and impact. One Department of Labor study, published in 1923, compiled results from questionnaires and interviews to form a comprehensive picture of children working in the beet fields of Colorado and Michigan. It uncovered that 20–25 percent of six-year-olds, 60 percent of eight-year-olds, and almost all ten-year-olds worked in the fields. During thinning season, children's average workdays ranged from nine to fourteen hours. While declining daylight hours reduced labor during harvest, the report stated that the typical child pulled and topped from twelve to fifteen tons of beets per day.[51]

The economics and agroecology of the beet sugar industry bore physical marks on growing bodies. Repetitive standing, stooping, and crawling targeted knees, ankles, and lower backs, while swinging hoes and slicing knives placed pressure on elbows, shoulders, and the developing muscles of the upper body. Multiple investigators noted some form of arrested development and physical pain stemming from beet labor. Winged scapulae, characterized by an abnormal protrusion of the shoulder blade, was one of the most common ailments. Inves-

Fig. 13. Sugar beet knife used by Gilbert Barela of Greeley during the 1950s. Photo by author.

tigators noted that it was caused by repetitive bending and cutting motions. In physical examinations, they also cited a preponderance of cuts and bruises, cracked and bleeding hands, and an inability to stand up straight. When interviewed, children often described persistent physical pains in necks, backs, and joints. One physician characterized the ailments as resembling rheumatism. Oral histories of beetworkers and formal investigations both highlighted the tensions between physical pain and work demands. Describing her sisters at the end of a long workday, Elva Treviño Hart observed that they simply laid down from exhaustion, seemingly unable to move and lacking any desire to replenish their bodies with needed food. During the night, it was common for sleep to compete with the physical pain inflicted by beet labor. Comparing the labors of children in manufacturing

Fig. 14. Seven-year-old Alex Reiber slicing the top of a sugar beet at harvest near Sterling. It was common for Alex and his peers to accidentally hook their knees while topping. Photo by Lewis Hine, Records of the National Child Labor Committee, Library of Congress, Prints and Photographs Division.

and beet fields, investigators concluded: "Beets are harder work than working in a steel mill [since] children don't get any fresh air as they have to lie in the dust and crawl on their knees all day."[52]

Statements such as these were calibrated to address a pervasive myth that enabled the contract labor system employed by Great Western Sugar—and throughout industrialized agriculture—to evade public scrutiny into the 1920s. While journalists and progressive reformers during the early twentieth century highlighted the injustice of children laboring for long hours in dangerous conditions in factories and sweatshops, they generally ignored analogous conditions in agriculture. The public presumed that child labor in the fields was a function of family-owned farms whereby work supported healthful activity for youth, exposing them to nature and imbuing them with values such as thrift and hard work—all of which were considered training for productive adulthood. Consequently, efforts to document the role of children within the contract labor system in regions such as the Piedmont aimed not only to expose injustices but to erode the stubborn belief that most agricultural labor was performed by families on their own small farms.[53] The Piedmont beet sugar industry provided evidence that child labor in the fields resembled factory labor, not family farming.

Children carried the marks of sugar beets with them into the classroom. The labor demands of thinning and harvest and the nature of beet contracts encouraged families to pull children out of school during harvest and thinning season. One 1916 study examined over thirty school districts on the Piedmont and found that the children of beet workers missed thirty-three out of the first sixty days of school. When these children returned to the classroom, they not only lagged behind their classmates but they were often underweight, malnourished, and had trouble focusing on schoolwork. For children of Mexican nationals, lack of proficiency in English presented another barrier and some schools prohibited the speaking of Spanish in the classroom. Teachers, administrators, and school boards commonly responded by stereotyping migrant children as lazy and incapable, or simply reacted in frustration at the enormity of the teaching task. Consequently, they further stigmatized children by holding them back. According to one study, as many as 80 percent of the children of beetworkers during the 1920s were behind grade level by one to six years.[54]

While Piedmont educators and the beet sugar industry acknowledged the practical and ethical quagmire presented by teaching the children of beetworkers, most disavowed responsibility and placed blame elsewhere. During the 1920s, Colorado law required work per-

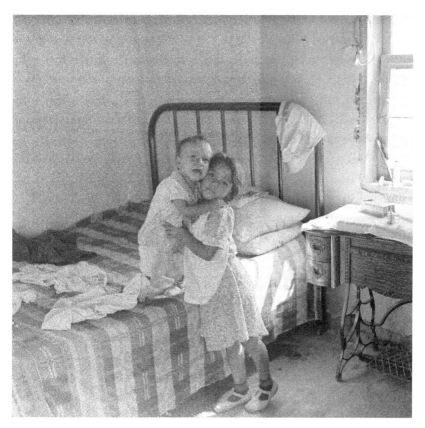

Fig. 15. A six-year-old girl taking care of her baby brother while her mother "works outside somewhere." This 1938 photo was taken in a Great Western Sugar *colonia*, illustrating a principal reason that children of beetworkers often missed school. Photo by Jack Allison, Farm Security Administration, Office of War Information photograph collection, Library of Congress, Prints and Photographs Division.

mits for children under fourteen and mandated that children attend school during the academic year from ages seven to fourteen. Those laws were rarely enforced. While beetworker families varied in their approach to sending their children to school, they justified removing children from school based on economic necessity. Growers argued that, if accomplishing the work required child labor, then families should contract for fewer acres. Great Western Sugar acknowledged the problem but argued that laborers signed contracts with growers, not the company, ignoring the fact that the company and the growers annually constructed the hand labor contracts employed throughout the region. School officials responded in a variety of ways. Bald racism was common. Officials argued that Mexican children were "animals" who could not be educated, or simply stated that Mexicans lacked enough initiative to "help themselves." Others simply resigned themselves to the reality of the situation, stating that the "heavy burden" of educating migrant children was too great, implying that such a burden precluded enforcement of attendance law. Elected school officials revealed the political consequences of demanding school attendance. In 1929 one superintendent claimed that he would lose the next election if he enforced school attendance.[55]

That school official came closer than most to acknowledging a reality that few wanted to admit publicly. Child labor and its educational consequences were functions of the economics of beet labor and the agroecology of sugar beets themselves. Great Western Sugar—and the entire beet sugar industry—recognized from its inception the need for a critical mass of hand laborers. Since the nature of beet seeds precluded rapid mechanization, the company and its growers solved that problem by hiring entire families of marginalized laborers to supply the physical demands of a crop that required hand labor. Child labor, and all of the physical and psychosocial consequences attached to it, was inseparable from Great Western's business model and embedded into the structure of Piedmont institutions such as the school system. Tackling child labor would require either mechanizing the industry or fundamental alterations in Great Western's business model—perhaps both.

The imprint of sugar beets on workers' bodies followed them from fields and classrooms into their homes. During the mid-1920s, the living arrangements of contract laborers and their families during the crop season varied considerably. Approximately half lived in the same beet shacks inhabited by their German Russian predecessors. Growers placed outhouses nearby and contracted with local municipalities to deliver the season's water into open-air cisterns. Laborers hauled all household water by hand. Families with longer tenure on the Piedmont commonly lived in one of two types of communities. About half dwelled in established, segregated neighborhoods in the nearest town. Most of the remaining laborers owned no-frills adobe homes built with sweat labor on land owned by Great Western Sugar. These worker colonies, or *colonias*, were the brainchild of the sugar company, constructed to rationalize its labor force, reduce transiency, and minimize the costs of annual labor recruiting. By 1930 almost one-third of Great Western's hand laborers lived in colonias. Even among beetworkers who owned or rented residences, the majority still spent all or part of each season living in beet shacks as time and distance rendered daily commuting untenable.[56]

Beet shacks brought the nature of labor into the home. Water presents the clearest example. Early season moisture, intermittent thunderstorms, and flood irrigation exposed leaky roofs and rough-cut timber walls, pulling mosquitoes indoors that were attracted to large families in concentrated spaces. Further, the open-air nature of most cisterns—on farms, in colonias, and most beetworker neighborhoods—invited disease. Children were typically tasked with carrying water into homes in buckets. As the volume of cistern water varied over the course of the season—cisterns were typically refilled every six weeks on average—foreign objects accumulated in the structures. One beetworker, Saul Sanchez, was offered a visceral examination of this sanitary quagmire when a grower hired him to clean out his family's cistern at the end of the season. Sanchez vividly recalls numerous insects and worms swimming in his drinking water. Reflecting on why his family did not get sick, he concluded, "We were used to those impurities, our bodies immune to poisons, as if our daily contact with the earth made us resistant to whatever crud we consumed." According to Sanchez, the nature of sugar beets could be measured in the bloodstreams and immune systems of its labor force. Additionally, beet soils and the transitory nature of the labor force brought the fields into the home. Since builders rarely designed dwellings with adequate ventilation in mind, families engaged the tradeoffs inherent in opening unscreened windows and doors to gain fresh air. Open doors alternately welcomed mosquitoes or dust, as well as the occasional snake. No amount of weatherproofing could resist other invaders. Bedbugs thrived where migratory labor and cramped spaces were common, as these pernicious insects fed on human blood and attached themselves to the clothing and luggage of migrant laborers. Sanchez recalls cycles of fitful sleep punctuated by bedbug bites and fruitless attempts to eradicate insect invasions. To solve the bedbug onslaught, some placed their bedposts in buckets filled with kerosene. Their stories demonstrate how the nature of fieldwork extended into the homes of laborers.[57]

Beet labor highlighted gendered burdens since it added to the labor of women and limited their ability to carry out traditional roles in family units. Women were responsible for keeping damp and mud-stained clothing dry and serviceable. They were up in the morning before daylight to make the meals that would carry families through the arduous workday, repeating the same task when tired bodies returned home. Since contracted laborers were not paid until the end of the season, it was common for them to obtain food in the spring through credit supplied by Great Western. Provisions were standardized based on cost and assumptions about the diets of Mexican families: "flour, Mexican beans, potatoes, salt pork, sugar, coffee, vermicelli, tobacco, canned tomatoes, chili peppers, spaghetti, white corn meal, matches, salt, pepper, lard, soap, and baking powder."[58] In her memoir, Elva Treviño Hart, the child of beet workers, stated that dinners consisted of "fried potatoes and frijoles guisados . . . the same as always."[59] The limited food choices functioned to enhance patriarchy. While living in beet shacks, laborers generally had little or no access to lands on which they could grow their own crops or maintain animals for food, such as chickens. These were tasks performed by women and children that contributed to the economy and health of the family and resembled the labors performed in the locales from which they migrated. The labor system perpetuated by Great Western and its growers demanded that women perform work typically assigned to men, while eroding their power over food sources. Sugar beets demanded women's labor and constricted their place within family economic units.[60]

Contract Labor and Environmental Risk

As sugar beet laborers bore the stamp of Piedmont agroecology in every aspect of their lived experience, Great Western Sugar and its growers placed the burdens of environmental risk on their shoulders as well. In April, when workers signed up to thin, weed, and harvest a specified number of acres, the company and its growers protected their economic interests by deploying contract language that insured them against nature's idiosyncrasies. Contracts stated that when any phase of cultivation was not completed on time, growers could hire extra labor to complete unfinished tasks, charging the cost of such work against the wages of the contract laborer.[61] Consequently, factors such as mud and frozen ground were not just environmental hardships; they could be economic ones. And weather was not the only environmental unknown for laborers. If workers did not already have a previous relationship with a grower, there was little way for them to know what kind of care the contracted acreage had received. Poor cultivation techniques and improper soil maintenance slowed thinning and weeding. But the greatest risk assumed by hand laborers was at harvest, when Great Western and its growers commonly sought to cheat nature by leaving beets in the ground late into October when the danger of a hard frost increased. When the weather cooperated, the gamble paid economic dividends to growers and Great Western. When it did not, the corporation and its growers often scrambled to hire extra workers, offloading the costs onto contract laborers who played no part in the decision-making process. While nature could not be rationalized, some of its financial burdens could be.

No year illustrates this better than 1929. In that year, Colorado growers contracted to plant nearly 250,000 acres of sugar beets—the highest in state history—with the majority grown on the Piedmont. This, combined with a decrease in the supply of resident laborers due to union protest, led Great Western's chief labor recruiter, C. V. Maddux, to seek more workers in New Mexico, Texas, and Mexico. Hand laborers that year agreed to cultivate fields at an average rate of twenty-three dollars per acre, plus bonuses for high harvests.[62] When warm weather held out in early October, Great Western and its growers delayed harvest in the hopes of reaping windfall profits. Meanwhile, laborers, anxious to complete the harvest and get paid for the season, openly questioned the delay.[63] When Great Western finally gave the green light to pull and top beets, the weather quickly turned. Alternating periods of rain and snow resulted in muddy fields, miserable workers, and a halting harvest. Then the weather turned frigid enough to freeze unharvested beets into the earth. Excavating deeply rooted beets slowed the harvest to a snail's pace. The cold temperatures generated angst since frozen beets lose sugar content. So, growers scrambled to hire whatever additional labor they could find, even releasing some prisoners from local jails to work in the harvest. The environmental and economic costs fell heavily on Mexican laborers as their anticipated wages disappeared with each new hire. One hand laborer stated that he was supposed to receive $154 for harvesting fourteen acres, yet he had to pay out $70 for additional labor and he still had three acres remaining before getting paid.[64]

Due to what one Piedmont reformer called "technical clauses" contained within beetworker contracts, the fallout from the greed and environmental hubris of Great Western and its growers fell most heavily on laborers. While some growers paid contract laborers throughout the season, most waited to pay wages in full until the end of the season, generally by November 15. With large quantities of beets still unharvested, most laborers were not compensated even their reduced wages until after they harvested the entirety of their contracted acreage. In some cases, this did not occur until January. Some laborers were not paid at all. In response, many filed wage claims with the Beet Workers Association, an affiliate of the American Federation of Labor. An incomplete list of those claims from forty-one laborers in one beet district documented $7,300 in unpaid wages. In Weld County, those who applied for relief from public agencies were often ignored, leaving many to seek the help of philanthropic organizations such as the Mexican Welfare Committee of the Knights of Columbus, a charitable arm of the Catholic Church. Many of those leaving the Piedmont for the winter resided seasonally in Denver, where relief funds through the city and Catholic charities quickly dissipated. As for Mexican nationals who possessed the least amount of recourse against their employers, some were faced with the choice of returning to Mexico without wages or pursuing uncertain Piedmont wage claims. This quandary was especially acute since most lived on the properties of the same growers who refused to pay their wages. After being evicted from those homes, some then turned to the Mexican Consulate for financial assistance to return to Mexico. At the end of the 1929 season, the consulate provided aid to 325 Piedmont families returning home. While Great Western and its growers paid for their 1929 harvest-time decisions in the form of reduced beet tonnage and fewer sugar sacks, they rationalized their decisions through contracts and practices that placed environmental risks on the shoulders of laborers.[65]

Piedmont agroecology was embedded in the bodies of laborers even as it was woven into the fabric of global capitalism, where Great Western and its growers sought competitive advantages in sugar markets by exploiting nature's energies for the lowest cost. Contract workers were a critical component within the larger set of energies employed to make sugar. Laborers possessed an intimate knowledge of sugar beet cultivation—buried deep in muscle memory and tuned to the particularities of each farm. Just as workers were inseparable from the farms they labored on, so the sugar beet landscape imprinted itself on the workers. Sugar beet agroecology worked its way into the crevices of their homes and into their food. It delimited educational opportunities for children and both informed and reinforced racial classifications. Deep knowledge of sugar beet cultivation also empowered laborers to negotiate for better housing, water, and working conditions. Despite this, Great Western and its growers wielded far greater power to manipulate human energies to produce beet sugar. To exploit labor further, growers often planted early and harvested late in the knowledge that, if weather slowed work, they could hire extra hands and charge costs against the wages of contract laborers. It was a system that supported efforts to cheat both nature and essential workers to produce a global commodity at the lowest cost.

Great Western Sugar's attempt to channel the efforts of contract laborers should also be understood as part of the company's larger aim to find cheap and efficient energy. This presented a stark and ironic set of divergences. In its relationship with growers, Great Western recognized that the most prudent route to sugar abundance came from a long-term investment in Piedmont soils. This required meticulous attention to crop rotations, beet byproducts, and fattening livestock. Profits, for growers and the sugar company alike, were based on a shared dependency on local resources. By contrast, cultivation energies were rooted in temporality and inequality. Since the nature of beet seeds precluded rapid mechanization and a critical mass of cheap labor did not exist on the Piedmont, the company and its growers exploited large families of seasonal migrant workers who existed on society's margins. Just as the company and its growers tried to squeeze every last ounce of sugar out of its beets, so they conspired to profit by squeezing the energies of laborers. Thus, the great irony of the Piedmont beet industry during the early twentieth century is that it sustained healthy soils by abusing human bodies.

Despite this, labor-saving technologies that included machines, chemicals, and single germ seeds were always in Great Western's purview. Like its industrial peers, the company leaned heavily on research subsidies to realize its goals. Scientists from various branches of the USDA and researchers from CAC and other land grant universities eventually enabled Great Western Sugar, and the entire beet sugar industry, to replace migrant muscle with machines and chemicals. Even as families of laborers contributed energy and knowledge to the beet sugar industry, the industry conspired with state-sponsored researchers to replace them. The research and relationships established by these agri-state scientists would transform human relationships and Piedmont agroecology in the years to come.

4 Piedmont Sugar and the State of Science

> What we call Man's power over Nature turns out to be a power exercised by some men over other men with Nature as its instrument.
>
> —C. S. LEWIS

In March 1899, researchers at Colorado Agricultural College's (CAC) Experiment Station devoted several pages of print space in its annual bulletin to describing the results of a sugar beet growing contest during the previous season. Its sponsors, including the Denver Chamber of Commerce, the city of Loveland, and several Piedmont counties, hoped to demonstrate the viability of a Piedmont beet sugar industry by awarding prizes for quantity of beets harvested, weight, and sugar concentration. The college's resident cattle expert, Wells Cooke, shared its sponsors' goals. Interpreting the results of the contest in an effusive light, Cooke argued that "the results of 1898 are so conclusive, that we may feel justified in saying that Colorado can raise as good sugar beets and as large crops of beets as any place in the world." Cooke's new colleague and promotional ally, chemist William Headden, joined in the chorus stating that "no place where a factory is now in operation presents advantages equal to those possessed by any one of half a dozen localities in Colorado."[1] When it came to beet sugar, college scientists, business interests, and industry promoters made common cause.

Scientists at land grant colleges such as CAC were not the only agristate researchers who clamored for a domestic beet sugar industry. As early as 1883, the U.S. Department of Agriculture (USDA) sponsored beet cultivation research, identifying regions where beets could be grown, and gathering information on constructing and operating factories for sucrose extraction. Harvey Wiley, head of the Bureau of Chemistry was especially keen on locating a source of sugar that could be grown in the temperate climes that characterized much of the United States.[2] Wiley, Cooke, and Headden are early examples of agri-state scientists who used their state-sponsored positions to promote the beet sugar industry. They believed that commercially grown beets required both significant private investment and a critical mass of public research scaffolding, and they were prepared to provide the latter. In their minds, no clear distinction existed between supporting sugar beet cultivation and forming alliances with the industrial entrepreneurs who emerged to lead it. Theirs was an implied mandate to do both.

When it came to the beet sugar industry in the American West, agri-state researchers at the USDA and land grant colleges understood that their best efforts could founder without proper attention to irrigation. This was the provenance of their engineering peers. Developing and maintaining the complex infrastructure that harnessed, stored, and redirected the South Platte watershed called for the kind of engineering expertise that was also concentrated within the USDA and CAC. Irrigation engineers designed infrastructure, shored up leaky reservoirs, developed inventions for efficient measurement and delivery of water, and acted as consultants for farmers and agribusiness. Their work impacted water's availability, distribution, and economic value, all of which bore a direct relationship with land use decisions. Their efforts, however, diverged from agri-state colleagues since, unlike peers in agronomy; chemistry; and biology, irrigation engineers rarely interacted directly with corporate clients, such as Great Western Sugar, or distant manufacturers of chemicals and fertilizers. The clearest beneficiaries were farmers with vested rights in dozens of mutual irrigation companies throughout the region. Yet, despite this difference, the agri-state scientists who engineered the Piedmont's water infrastructure during the early twentieth century were most effective and influential when they calibrated their research to meet user demands. Thus, like their agronomic peers, irrigation experts charted courses that mapped onto clientele agendas.

In that sense, Piedmont agriculture inverts and diverges from arguments about the relationship between the federal government and the West in the early twentieth century.[3] One of those arguments regards the region's deep reliance on government largesse for its development. While federal funding was important, state-sponsored science depended on the beet sugar industry and water users to set agendas, rarely charting an independent institutional course. Neither did federal and state resources and expertise define the relationship. Despite growing agency budgets, beet sugar manufacturers always possessed capital far beyond anything agri-state agencies brought to the table. When it came to water, Piedmont farmers never wavered in their demand for more, and irrigation engineers were effective to the degree that they tuned their research agendas to the water abundance frequency.[4] During the early twentieth century, agri-state scientists allowed industry and water users to set research directions, and it was in that subordinate and decentralized context that they impacted Piedmont sugar beet agriculture.

These evolving institutional relationships had immediate and long-term consequences on Piedmont agroecology. During the early twentieth century, efforts by agri-state scientists supported rationalizing the existing system of mixed farming that sustained regional soils and sought secure water supplies. In the process, agri-state scientists networked with industry and powerful farmers, rarely charting independent research courses. Over time, this reduced their research independence and eroded their ability to direct and, in some cases, critique the region's farming. Moreover, while it may not have been apparent during the first three decades of the twentieth century, the networks established by agri-state scientists eventually cleared the path for Piedmont farmers to abandon long-established farm practices. This chapter details the dynamics by which agri-state scientists developed research and relational networks with farmers and industry during the first three decades of the twentieth century, anticipating how those evolving relationships delimited future efforts.[5]

Agri-State Establishes a Piedmont Presence

Federal support for sugar beet research came from USDA scientists working in the midst of growth and philosophical reorientation. During the early twentieth century—a period characterized by federal expansion—no department grew more rapidly than the USDA. During a fifteen-year period from 1897–1912, USDA employment increased from 2,440 employees to almost 14,000, and its budget multiplied sixfold while it became the most dynamic agency in the federal government. New agency leadership prioritized scientific expertise over bureaucratic experience and agency heads, such as Beverley Galloway of the Bureau of Plant Industry (BPI), actively recruited scientists from research universities to work on the pressing agricultural problems of the day. Scientific expertise enabled agency and division heads to act independently of federal oversight and build coalitions. Desiring the broadest impact, USDA employees—from agency heads to field scientists—sought out heavily capitalized large farmers and industry leaders who utilized modern production methods, enabling USDA scientists to broadcast the results of their work to a wide audience. In Colorado that research agenda led USDA scientists directly to the sugar companies.[6] While this supported industry imperatives, it shielded researchers from the majority of growers—most of whom farmed modest acreage—and certainly from the hand laborers who performed the lion's share of the work.

Experiment station researchers at land grant universities, such as CAC, largely shared the prerogatives of their USDA cohort. Initially, this presented obstacles since farmers applied pressure on state agricultural boards who were responsible for providing direction to land grant colleges. That control loosened when Congress passed the Adams Act in 1906 to provide experiment stations with funds earmarked for theoretical research. This change allowed agricultural scientists at land grant colleges to develop independent research agendas. It also freed them up to collaborate with scientists working on similar research questions. Where that research placed them in the same fields as USDA scientists, researchers generally made common cause around shared agendas, trumping administrative rivalry. Consequently, research in Piedmont beet fields increasingly involved scientists from federal and state agencies who crossed institutional boundaries both within individual projects and in their career trajectories.[7]

Two years after Great Western Sugar built its first Colorado factory in 1901, the USDA's BPI formed the Division of Sugar Plant Investigations (SPI) in 1903 to support the growing beet and cane sugar industries. Headed by plant pathologist Charles Townsend, scientists at SPI were guided by a broad USDA mandate to enlarge domestic sugar production in beet-growing regions while expanding sugar beet agriculture onto new lands. The most promising regions were in the American West.[8] At its inception, SPI fieldworkers recognized that their research depended on industry scientists and growers with more hands-on experience. Monogerm seeds provide a poignant example and the most valuable frontier for research. Single germ seeds planted at uniform distances could drastically reduce the need for hand labor since they could synchronize with mechanized cultivators, eliminating weeds at precise intervals. This depended on locating and breeding rare strains of single germ seeds and likely required coordination with beet growers worldwide. SPI lacked the resources for such an undertaking. Consequently, SPI scientists focused instead on breeding seeds adapted to the various climates and soils in the beet growing regions of the United States. At the time, the beet sugar industry imported European seed, bred for European microclimates. Attempting to jumpstart its research, SPI researchers sent out questionnaires aimed at discovering which imported seeds proved most successful on American soils.[9] They asked sugar companies for a list of imported beet varieties sown on its growers' land, correlated with the yields and percentages of sugar produced by each seed variety. Townsend emphasized that complete responses would propel better seed selection, pointing out that resulting increases in sugar would generate rising profits for each factory with virtually no increase in manufacturing cost.[10]

Responses from beet sugar companies revealed several inefficiencies that hindered SPI research goals. Great Western Sugar submitted tables from its growers displaying acreage and yields for consecutive years. But growers generally did not record data explaining why some years and certain fields were more productive than others. One grower, William Stanley of Eaton, meticulously charted the acreage he planted in beets each year, tons harvested, payoffs by Great Western, and net profits. Though Stanley could detail which varieties of seeds he planted, he could not correlate them with yields or sugar content.[11] Great Western also bore some responsibility for spotty tabulation. The company was responsible for purchasing seeds—the majority being from various seed houses in Germany—and then selling them at cost to its growers. But Great Western made no attempt to track them from point of sale to harvest. That responsibility lay with the grower. In pointing out these complications while also trying to be helpful, W. A. Dixon, secretary for Great Western, included a pound of refined sugar from each of its factories along with the company's incomplete questionnaires from 1904.[12] The spotty questionnaires made it clear that SPI's attempts at data gathering required a more hands-on approach.

That direct approach involved developing its own seed research, in growers' fields and at industry and agricultural college experiment stations. With the goal of materially increasing the sugar yield locked in each seed, SPI conducted research at agricultural college experiment stations in Colorado, New York, Michigan, Utah, and Washington, as well as on the USDA's own land in Arlington, Virginia. Great Western placed SPI researchers into contact with large growers who volunteered land for these experiments. The results, according to Townsend and other SPI researchers, was that by 1908, they had already achieved sugar yields on a commercial basis higher than those of the major European growers.[13]

Yet the beet sugar industry remained reliant on imported seed. Why? The answers were found in science and the free market. An initial sugar beet seed crop required two years. The roots formed in the first year, and seed growers left them in storage over the winter. Then they replanted the roots the following year. The stalks that formed from these roots yielded the seed germs. Seed growers selected the best germs from that crop and then repeated the process. Typically, a commercially viable crop of seeds required six years. Through trial and error, SPI scientists discovered that commercial seed production required a milder year-round climate and warmer soils than were available in Michigan and Colorado, the leading sugar-producing states. Consequently, they conducted research in Arizona and the temperate areas of western Washington and Oregon. Though these experiments resulted in commercial-quality seed, production could not meet grower demands. In 1909 Townsend lamented that only 5 percent of all seed needs were being met domestically. He hoped in vain that independent seed companies would emerge to take on the challenge, but they were unwilling to accept the risks and delay associated with commercial production.[14] Further, beet sugar refiners were loath to experiment because the domestic seed payoff would not be realized for several years, if at all. In the meantime, foreign seeds were relatively inexpensive, even if they were not adapted to the beet growing regions of the United States.

Judged by its own standards, SPI's seed work failed during its first decade. In 1901 Townsend argued that breeding beet seeds suitable to the varying climates of the United States would result in out-competing European growers. Yet, in 1914 the United States remained reliant on imported seed. Lamenting the commercial consequences should trade with foreign seed houses be cut off, Townsend opined that "our sugar beet industry in America is dependent upon no inter-ference with Germany and other sections of Europe."[15] If industry adoption of seeds bred for regionally specific climates and soils was the primary metric for SPI's success, then its work was a qualified fail-ure. While companies such as Great Western were eager to cooperate, they were not willing to invest in seed enterprises based primarily on agri-state research. SPI's early efforts showed that it could serve industry but could not direct it.

Yet, industry leadership is a misleading and limited metric for eval-uating SPI's early efforts. Measured by volume of research and networking, SPI wielded some clout by 1910. Its researchers moved from simply gathering industry and experiment station data to partnering in and overseeing research projects. By 1909 SPI published thirty-eight bulletins and circulars on subjects as wide-ranging as fertilizer, crop rotations, insecticides, the use of beet byproducts, and seeds bred for American soils and climates. These publications were distributed either directly to growers or indirectly through beet sugar companies and agricultural experiment stations. Moreover, as the beet sugar industry grew, SPI's budget and influence grew astride.[16] By 1910 SPI employed two pathologists, a plant physiologist, four fieldworkers, and several lab assistants at the same time that it developed working relationships at every level of the beet sugar industry.[17] In response, company leaders began requesting help with specific problems. Since aiding industry was inherent in its mission, SPI was happy to oblige.

Industry Sets the Agenda: The Southern Colorado Piedmont Example

One of the first industry requests for USDA aid came from William Wiley, president of Holly Sugar. In 1910 Wiley operated two beet sugar factories on the Southern Colorado Piedmont, in the Arkansas River Valley towns of Holly and Swink. Though sugar beet agriculture on the Southern Colorado Piedmont was less developed than its northern neighbor, it shared characteristics that made agri-state research performed in one region relevant to the other. The Arkansas River was the Southern Piedmont's corollary to the South Platte River, providing nearly all of the irrigation necessary to grow beets. Both regions possessed plenty of sun during the growing season, cool evenings, and similar soil profiles. Like its northern neighbor, the Southern Piedmont possessed ample rail access.[18] Holly Sugar was largely successful during its initial years of operation, as growers harvested bumper crops and sucrose yields per acre exceeded those achieved by Great Western Sugar. However, in 1910 yields plummeted and average sugar content within beets fell by 50 percent. Disaffected growers threatened to discontinue cultivating beets and blamed the company for failing to adequately address their woes.

Fig. 16. Unloading sugar beets at the Swink factory, ca. 1906–10. Growers in the Swink area were among those hit hard when yields plummeted in 1910. Photo by H. H. Buckwalter, History Colorado, Denver, Colorado. Used by permission.

Initially Wiley brought his concerns to the CAC Experiment Station, but he became impatient after its researchers concluded that excess soil nitrates were the problem but did not immediately offer a solution.[19] So, Wiley sought out SPI. Seeing an opportunity to broaden the scope of its work, SPI's head pathologist reached out to William Spillman, founder and agriculturalist in charge of USDA's Office of Farm Management (OFM), requesting that he complete an assessment of the sugar beet conundrum in the Arkansas Valley. Spillman agreed.[20]

Spillman was an ideal choice. As a plant scientist by training, he focused his energies on applying plant pathology research to the economic problems of farmers. Additionally, Spillman emphasized collaboration with experiment station researchers at agricultural colleges and cooperation with industry leaders. Previously, Spillman initiated fifteen cooperative projects between USDA scientists and

experiment stations. Having placed several college-trained agriculturalists in northern and western states, Spillman directed specialists to
disseminate relevant agricultural research to farmers through meetings, bulletins, and farm demonstrations. He employed this model of
education and cooperation between agri-state researchers, industry,
and farmers in the Arkansas Valley.[21]

From his initial 1910 investigations, Spillman developed several
explanations for Holly's woes. Disagreeing with CAC findings, he
concluded that nitrogen deficiency, not excess, led to low sugar yields.
Offering a historical explanation, Spillman argued that for several
decades prior to the advent of the local beet sugar industry, overgrazing wore the natural grasses thin, grasses that fixed life-giving
nitrogen in the soil. Thus, when farmers initially tilled the soil and
planted beets, sufficient nitrogen existed only for a few years of high
yields before the soil crashed.[22] He also suggested that over-irrigation
had saturated some soils beyond use and that lack of careful attention
to cultivation stunted beet growth. Additionally, Spillman encountered an epidemic of leaf spot (*Cercospora*) on sugar beet plants, an
ailment that sapped plant energies during summer months when
healthy plants grew rapidly and concentrated sugar. Finally, in an
indirect nod to the need to accelerate production of domestic beet
seed bred for local climates, Spillman placed some blame on German
suppliers who might be "sending over a lot of poor seed."[23]

Improving productivity required further research, but it also
entailed careful politics. If industry or experiment station scientists
perceived federal researchers usurping their investigations or dictating
the direction of their business, SPI could be shut out of future projects.
Consequently, Spillman emphasized that his researchers should make
every effort to befriend Holly company executives, agriculturalists,
and irrigation experts. He further advised that, even if agri-state scientists were not impressed by sugar company research, they should
remain deferential and open to industry advice.[24] Spillman's influence
was also evident in relationships with CAC. In conversations with
CAC Experiment Station director Louis Carpenter, USDA pomologist
Harold Powell assured Carpenter that the USDA would not inter-

fere with its soil studies. Further, Powell explained that the project was "part of a broad investigation of the fundamental problems in the breeding, pathology, and nutrition of sugar beets, which we are carrying on in several states." In other words, the work was not an attempt to supplant CAC's authority in Colorado but part of an effort to support beet sugar industry success. Powell additionally promised that all findings would be shared across agency lines.[25] According to Powell's logic, CAC's research would not only continue without interference but, through agri-state collaboration, it would achieve greater relevance and a wider audience. The reasoning Spillman and his colleagues employed suggested that federal inclusion in the sugar beet research complex required that its scientists actively break down administrative and research barriers between itself, industry, and experiment stations while allowing industry partners to set agendas.

Spillman's efforts to attach agri-state researchers to industry needs extended beyond the physical and institutional boundaries of Southern Colorado and Holly Sugar. At the start of the project, Spillman procured the consulting services of Charles Townsend.[26] This presents as a minor point except that Townsend left his official post as director of SPI in 1909 to take a research position at the U.S. Land and Sugar Company in Garden City, Kansas. As a beet sugar manufacturer and a company whose downstream growers employed the same Arkansas River waters as Southern Colorado farmers, U.S. Land and Sugar was one of Holly's competitors. Though Townsend never received a paycheck from Holly Sugar, he was indirectly offering them his services. Townsend's position suggested that while agri-state researchers aided individual beet sugar companies, their allegiance was to an entire industry. As beet sugar companies sought profit from federally subsidized research, state-sponsored scientists employed individual projects to network across an entire industry.

Townsend's role illustrates why that had become increasingly possible. He was one of a growing network of federal scientists who possessed unique knowledge and skills desired by industry. After heading SPI for six years, Townsend was an expert in all phases of sugar beet cultivation and had established relationships with each of the nation's refiners.[27] That knowledge and experience in public employ was coveted by private industry, usually at a higher pay grade. But Townsend's unique set of skills enabled him to maintain agri-state ties and conduct research that was not limited by the immediate interests of his employers. Townsend was the first of many SPI scientists to obtain research positions in the beet sugar industry. His example and the work of researchers in the Arkansas Valley demonstrate how the state accomplished its mandate to support industry—not through centralizing power, but through dispersing expertise across agency lines, collaborating with land grant colleges, and cultivating and maintaining scientific networks that transcended state and industry boundaries.

The work of sugar beet researchers in the Arkansas River Valley connected Holly Sugar's production with larger regional and national goals. The USDA work for Holly employed twenty-nine workers on 425 acres. Their research covered the entire gamut of sugar beet farming: irrigation, cultivation, the use of fertilizers and insecticides, crop rotation, and harvesting. Their work even delved into some of the social complications of beet farming such as tenancy and migratory labor.[28] In short, state efforts were a microcosm of the needs of an entire industry. Additionally, the research offered encouragement—in the form of proven research and ongoing support—to regions that were flirting with sugar beet agriculture.[29] Successful research in Southern Colorado also translated to more domestic sugar and less reliance on imports from far-flung locales such as Cuba and the Philippines. More than that, it declared, in unveiled terms, that the federal government possessed funds and expertise to offer technical and financial assistance to an industry of growing importance to national interests. That value was about to become more apparent.

Agri-State Goes to War

The onset of World War I in Europe sent shock waves through international sugar markets. In 1914 beet sugar provided nearly one-half of the world's sugar, while its largest producers, Germany, Austria-Hungary, Russia, and France warred with each other. The conflict decimated their respective industries and squeezed sugar shipments.

On the surface, this was not a crisis for the United States since none of the warring nations were critical suppliers. However, a closer look reveals concerns over sugar security. Prior to the war, importers such as Great Britain relied heavily on sugar from belligerents. That sugar was no longer available. Further, during the conflict, Britain could not spare ships for the long voyage to obtain supplies from producers in the Pacific. Consequently, it turned to Cuba, which supplied the United States with nearly half of its sugar. As demand outpaced production, prices rose. Compounding the sugar squeeze, Germany and Austria-Hungary supplied the majority of seeds for the beet sugar industry. This concern became a crisis in 1917, when the United States declared war on those nations, allying with Great Britain, its chief competitor for the world's dwindling sugar stocks.[30]

Both nations viewed sugar as an essential wartime foodstuff. Though sugar plays no part in replenishing the body's tissues, its primary wartime value was to produce energy, and no other commodity performed the same function at a reasonable cost. At 1917 prices, sugar supplied one hundred calories of energy for six-tenths of a penny. Sugar's quickly metabolized vigor provided the most significant source of calories for troops overseas.[31] Moreover, refined sugar was dense and easily packed, enabling simple transport during a war when cargo ships were sparse. Sugar was also an essential wartime consumer good. As Raymond Pearl, chief of the Statistical Division of the U.S. Food Administration (USFA), argued, when other foodstuffs were scarce—common in wartime—people instinctively reached for sugar because of the immediate energy it provided. Writing in 1918, he argued that the war emphasized this reality since significant sugar rationing among the combatants resulted in "more discomfort, discontent, and loss of morale than reduction in any other food." Pearl's rhetoric implied that the state was especially interested in regions that could produce sugar in quantity.[32]

The USFA and the beet sugar industry expressed the state's interest in sugar security through their wartime efforts. Formed in August 1917, the USFA made sugar a top priority. Concerned about market volatility, its officials signed voluntary agreements with all domestic beet sugar manufacturers to buy their sugar at fixed prices that would guarantee profits without alienating consumers. USFA officials then brokered negotiations between growers and industry that guaranteed growers a rate of ten dollars per ton of beets. They developed rules to limit consumption as well, employing a host of local food administrators to ensure that these rules were being followed. To guarantee wartime supply, USFA officials convinced President Woodrow Wilson to form a government-owned corporation, the Sugar Equalization Board, in 1918. Authorized to buy Cuba's entire sugar crop at fixed prices, the Sugar Board was also charged with equalizing wholesale prices for beet and cane sugar. Few agricultural commodities garnered as much state attention as sugar during World War I.[33]

Regulation of sugar during the war contributed to economic crises that highlighted ongoing state interest in the beet sugar industry. Motivated by a guaranteed buyer and strong wartime profits, Cuban cane sugar producers expanded their acreage following the conflict. American colonies in the Pacific, Hawaii, and the Philippines followed suit. In addition, by the mid-1920s the European beet sugar industry recovered, producing at almost prewar levels. Within a few years, sugar was in a state of marked overproduction. As the domestic beet sugar industry spread fears of financial collapse, Congress approved three separate sugar tariff increases from 1921 to 1930.[34] Congressional arguments resonated in the executive branch as evidenced by Calvin Coolidge's claim that removing tariff protection would "destroy our domestic industry," that a robust beet sugar industry was essential to domestic food independence, and that consumers were best served when plentiful beet sugar entered the market.[35]

Agri-state scientists working with the beet sugar industry understood that beet politics and the science of their cultivation were not mutually exclusive. Industry calls for economic protection reverberated in relationships with agri-state researchers. Prior to World War I, SPI invested much time and treasure into researching sugar beet agriculture and touting its benefits to regions of the country that had not embraced it. Its reasons were nationalist, ecological, and economic.

Armed with a strong tariff and a growing budget, SPI scientists argued that more acres in beets would result in reduced reliance on imported sugar. They further claimed that sugar beet agriculture, with its dependence on rotating crops, meticulous cultivation, and integration with the fattening of livestock, such as cattle and sheep, was healthy for the land and the long-term viability of farmers. Finally, SPI posited that sugar beet agriculture was profitable. However, after World War I, investors had little interest in opening new factories since market prices for commodities such as sugar beets were volatile and consumption plateaued. Moreover, with increased quantities of cheap cane sugar available from island producers such as Cuba and the Philippines, adding acreage and factories appeared risky. Finally, though the tariff provided a strong wall of protection, there was no telling when political pressures might cause it to crumble. Consequently, SPI and its network of scientists shifted away from stressing more acres and factories and toward greater efficiency on existing acreage.[36]

Prospects to engage the efficiency imperative were limited in Piedmont beet fields during the 1920s. Since the industry was years away from replacing contract workers with machines, labor costs offered little wiggle room. In fact, the cost of paying contract laborers in the 1920s increased slightly due to occasional worker shortages and emerging public scrutiny of industry labor practices.[37] Human cultivation energy would remain essential into the foreseeable future. Available water also presented limits. During the 1920s new agricultural lands on the Piedmont with the secure irrigation rights critical to growing beets were scarce. Short of a massive increase in water storage or more equitable water distribution, the remaining acreage in the region could only be used for dry land farming or stock grazing. More water could grow more beets, but it was not within sight. These limits resulted in a conservative approach to industry expansion. After a period of impressive growth from 1900 to 1914, Great Western Sugar built only three new factories prior to World War II.[38] For the company and its growers, the solution lay in doing more with existing soil, air, and water.

Product Testing for the Chemical Industry

This push for greater efficiency led SPI and beet sugar manufacturers to embrace the chemical industry. American chemical companies profited during World War I from federal contracts to provide compounds, including mercury, chlorine, formaldehyde, and ammonia—useful for concocting explosives and poisonous gases. In the wake of the war, chemical companies sought domestic markets for wartime surpluses. Agriculture represented a largely untapped outlet and possessed a support infrastructure capable of accommodating industry. During the 1920s it was common for agri-state scientists, regardless of employer, to conduct shared research at agricultural colleges, where sufficient land and the particularities of climate and soil enabled effective experimentation. Thus, when chemical companies sought product testing, they enlisted researchers at these state-sponsored institutions to conduct field trials.[39] To determine the efficacy of various chemical compounds on sugar beets, the chemical industry requested that SPI test several compounds. SPI agreed.

In 1925 several chemical companies, including DuPont and Bayer, developed compounds that included mercury, formaldehyde, and chlorine for a series of sugar beet experiments in Colorado and Utah. They instructed researchers to emulsify beet seeds with the compounds, referred to as seed treatments or disinfectants, aiming to kill disease-causing fungi and accelerate the germination process. During the 1920s sugar beet productivity on the Piedmont was hindered by *Cercospora*, or leaf spot, a disease characterized by small spots on foliage, resulting in withered plants, declining sugar content, and a loss of up to 40 percent in beet tonnage.[40] To conduct the experiments, SPI collaborated with American Beet Sugar's lead researcher, Anton Skuderna, in a Piedmont project that received assistance from CAC's regional experiment station. Skuderna hoped that emulsifying seeds with industry chemicals prior to planting would inoculate the plants against leaf spot and encourage faster germination, thereby enabling a longer growing season. SPI provided twenty-five strains of sugar beets for the experiments while CAC contributed land and researchers.

Seed disinfection is such a new science, relatively, that your customers have to be reminded of its advantages at every turn. If they buy only a 50c can, you've started a sales pyramid that will rise to unguessed heights.

The crop results are generally much better than anticipated. That makes for enthusiasm — and enthusiasm is contagious. And contagious enthusiasm makes for a big SEMESAN business.

Catalog reminders like these, when your customers are figuring out their seed requirements, will bring you a handsome return in SEMESAN sales.

CATALOG COPY SUGGESTION NO. 7

CATALOG COPY SUGGESTION NO. 8

Fig. 17. Ad for Dupont's Semesan Seed Disinfectant. Hagley Library, Wilmington, Delaware. Used by permission.

Though results were mixed, several of the sugar beet test plots exhibited mild improvements in germination rates, prompting Skuderna to pronounce them "outstanding," "worthwhile," and "indicative of promise."[41] These experiments, involving the same set of institutional actors, expanded over the next decade to include Dow and Shell's chlorine-based fumigants, and several ammonia-based fertilizers.

The chemical industry reaped rewards from its connections with state-sponsored researchers that extended beyond product field trials. Collaboration with scientific expertise generated product legitimacy. Consider DuPont's first successful seed disinfectant, hydroxymercurichlorophenol, or Semesan. As its chemical name suggests, Semesan synthesized both chlorine and mercury into the compound. After some initial testing, DuPont convinced SPI researchers, as well as experiment station scientists in Colorado, Michigan, and Utah, to include Semesan in their 1925 field trials. Slight increases in beet yields from Semesan-treated seeds resulted. Consequently, DuPont chose to manufacture the product commercially. In magazine ads and on product labels, DuPont claimed that Semesan propelled rapid and efficient germination of sturdy, early maturing crops, and bountiful harvests.

The company also quoted state-sponsored scientists to impress consumers. In one of its 1927 catalogs, DuPont claimed that Semesan had been "tested by agricultural colleges and experiment stations," and was "recommended by government experts, agricultural colleges, county agents and large growers in every section of the country." DuPont went on to quote farmers and extension agents who had participated in these experiments. One sugar beet farmer claimed, "The Semesan-treated beets were much stronger and healthier right up to harvest." Another extension agent exhorted, "Tests of Semesan on beets, sweet corn, peas, and tomatoes . . . have shown that the tendency (for fungal infection) is almost entirely eliminated." Since researchers were compelled to publish their findings, DuPont received free publicity in the form of USDA and experiment station journals and bulletins. By the 1930s chemical companies such as DuPont, and its seed treatment products, were fixtures in Piedmont agriculture.[42]

The drive for efficiency in the beet fields of Colorado and the West, alongside the expansion of the chemical industry, fostered an expanded state presence by making agri-state researchers conduits for industry field trials. This can easily be missed. Since chemical manufacturers were rarely present at field trials, their sponsorship only infrequently appears in the historical record. Further, when experiment stations recorded chemical names, they seldom cited manufacturers, instead using generic compound names, often because multiple companies branded their versions of the same compound. Without an industrial trail to follow, it's easy to conclude that these chemicals were concocted by agri-state scientists. The reality was just the opposite. Growing industry presence brought additional funding and opportunities to agri-state researchers. This enabled land grant colleges and USDA divisions such as SPI to hire scientists and undertake new projects aimed at making agriculture more productive. Consequently, industry field trials increasingly supplanted original research. So, while the quantity of agri-state research projects grew in Piedmont beet fields during the 1920s, presence was tethered to industry agendas.[43]

Politicizing Water: Efficiency versus Quantity

As with the beet sugar industry during the early twentieth century, the expertise to engineer quantity and efficiency into the system that delivered irrigation water to Piedmont farmers resided in agri-state scientists. The presence of the federal government in the development of water resources in the American West during the first half of the twentieth century is unmistakable. While historians have highlighted the role played by the Bureau of Reclamation in that process, a diverse collection of researchers from agencies outside of Reclamation manipulated the region's waters.[44] These scientists and engineers were employed within various USDA agencies or as collaborative researchers at land grant colleges. They designed irrigation infrastructure, shored up leaky reservoirs, developed inventions for efficient measurement and distribution of water, and acted as consultants for farmers and agribusiness. Their work impacted the amount of water available to farmers and the quantity and economic value of individual water rights, and it enabled farmers to correlate crop selection with predicted runoff from snowpack. Since the goal of these scientists and engineers was to squeeze every drop of economic value from existing watersheds, their work usually took place inside labs or at the site of canals, ditches, and local reservoirs. Their labors did not inspire the sort of visceral response effected by a massive Reclamation dam. Yet, evidence of their collective impact on the plumbing of Western agriculture abounds.

Irrigation engineers working on the Piedmont researched, designed, and built much of the infrastructure that underwrote the economy and agroecology of the region. By 1900 farmers and industrial operators such as Great Western Sugar had squeezed every last drop of usable water from the South Platte River watershed and still demanded more. Between 1900 and 1920, Colorado still vied with California for the nation's lead in irrigated acres, with the Piedmont gulping 40 percent of Colorado's total.[45] Where water flowed, so did money. The value of water rights for irrigation from the Cache la Poudre—the South Platte's largest tributary—averaged $400 apiece in 1880, and $4,500 in 1917. Lands with excellent water rights escalated notably. In 1922, irrigated land near Greeley averaged more than $300 per acre. Fifty years earlier, $300 could buy eighty acres of Union Colony land with drought-proof water rights. While various factors propelled the escalation, water was the principal one. According to a Fort Morgan newspaper, when local boosters and farmers combined forces to build and store water in nearby Jackson Lake in 1904, "application of the lake water had almost a magical effect upon land values. Land[s] that had been offered for $50 per acre the previous year doubled almost overnight and [. . .] continued to rise." Engineering, irrigating, and wealth accumulation were mutually constitutive processes.[46]

Where irrigation abounded, crop choices narrowed. As new water rights became scarce and the value of existing ones swelled, landowners and tenants increasingly prioritized high value cash crops that would make their irrigation water pay. The beet sugar industry underscored this reality since sugar beets were among the most water dependent

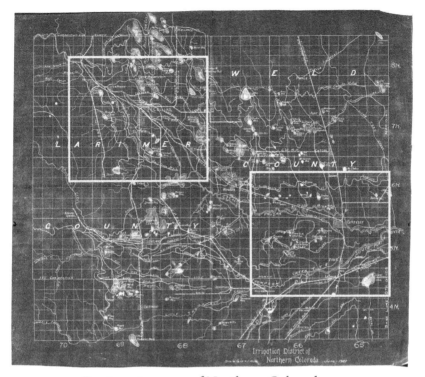

Engineering map of Northern Colorado

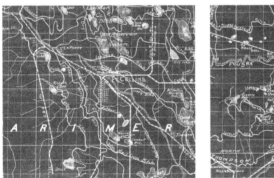

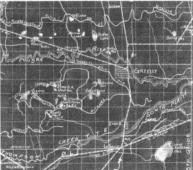

Detailed view Fort Collins   Detailed view Greeley

Fig. 18. Map of Piedmont irrigation infrastructure, 1921. The inset on the left highlights the Fort Collins/Loveland region. The inset on the right highlights the Greeley area. Map shows the density and interconnectedness of Piedmont irrigation infrastructure. Original map by Ralph Parshall, Irrigation Research Papers, csu Morgan Library Archives and Special Collections. Used by permission.

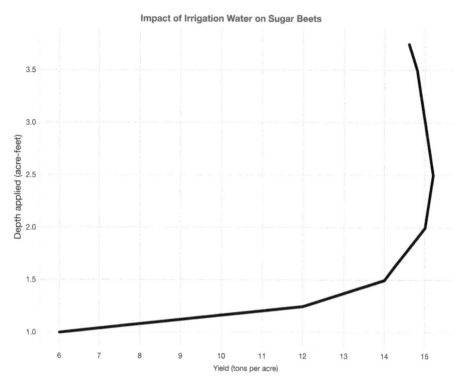

Fig. 19. Water correlated with sugar beet yields. Graphic created by Philip Riggs and adapted from Roger Hemphill, *Irrigation in Northern Colorado*, 1922.

crops in the region. Farmers with enough water to apply two acre-feet of water to each acre of beets during a season more than doubled the harvest of those who applied half that quantity. However, farmers tapping the Cache la Poudre River received an average of only 1.25 feet of water for each acre of irrigable land. Among other common crops, alfalfa was one of only two common cultivars requiring more water than sugar beets (potatoes were the other). As a cover crop that fixed nitrogen in the soil and fattened livestock, alfalfa was an essential culti-var. Full canals and ditches slaked the thirsts of the Piedmont crops that held the region's agroecology in place. Moreover, water security and quantity made the difference between abundance and financial ruin.[47]

While managing irrigation remained in the hands of the same farmer-owned mutual irrigation companies established during the nineteenth century, the expertise to distribute that water efficiently and equitably required engineering. During the early twentieth century, irrigation engineers working for CAC and the USDA's Bureau of Agricultural Engineering developed techniques for measuring water more effectively, shoring up leaky infrastructure, building water storage, and predicting annual water flow. No irrigation engineer played a more crucial role in this than Ralph Parshall.

During the 1920s and 1930s, Ralph Parshall developed measurement techniques that vastly increased the efficiency of water distribution in the South Platte Watershed. This provided a platform to influence some of the most critical water decisions of his time. Born in 1881 and raised on a Piedmont farm, Parshall entered CAC in 1899, majoring in civil engineering. His interest in hydraulics led him to focus on irrigation. After spending several years working at the Colorado State Engineer's Office and completing graduate work at the University of Chicago, Parshall was hired by CAC as an assistant professor of civil engineering in 1907. His formal employment with the university came to an end in 1913 when he took a position as an assistant irrigation engineer with the USDA's Bureau of Agricultural Engineering. In 1918 Parshall was promoted to senior irrigation engineer, a post he held for thirty years. Though officially a federal employee, much of Parshall's work employed CAC's facilities and its students. His career, like that of his agri-state peers aiding the beet sugar industry, illustrates the blurry and overlapping divisions between federal and state researchers and the agencies they worked for. Parshall planned and built hydraulics laboratories on the CAC campus, where dams and irrigation structures on the Piedmont and throughout the irrigated West were modeled and tested. These included Hoover and Grand Coulee Dams, as well as dams within the Tennessee Valley Authority and the Panama Canal Zone.[48] With his background in farming and irrigation engineering and connections to engineers at the state and federal levels, Parshall was equipped to diagnose the weaknesses within the Piedmont's irrigated agriculture and trained to develop technical solutions.

Like his agri-state colleagues researching beet sugar cultivation in the 1920s, Parshall viewed inefficiency as the most significant problem facing his Piedmont clientele. As Parshall surveyed how farmers irrigated their lands, he observed that they failed to manipulate water as they should, concluding that technocratic solutions were necessary in both the engineering and governance of the resource. In 1922 Parshall provided one of the first complete expositions of water seepage in the region. His research showed that 30 percent of all the water withdrawn from the watershed returned back to it downstream as a result of seepage.[49] In financial terms, Parshall calculated that this water could generate $2 million annually if efficiently applied to farms and fields. According to Parshall, Colorado farmers should be alarmed that so much of this water flowed out of the state when the system could accommodate more downstream reservoirs. Parshall measured water efficiency in quantities of crops as well as money earned. In advocating for two new dams on the Cache la Poudre River in 1925, Parshall's metric was sugar beets, stating that these dams would provide the necessary water to grow two additional tons of beets per acre. Elsewhere, Parshall complained that reservoirs were notoriously leaky, losing half of their water to seepage and evaporation, implying that farmers could have more secure water supplies if only they invested more money—through their mutual irrigation companies—in the modern engineering necessary for delivering this precious resource. If water draining from creaking structures presented one kind of inefficiency, then water that failed to drain presented another. Parshall opined that plenty of good farmland was out of production because farmers either applied too much water to their lands or water seeped onto their lands from neighboring farms. Better drainage could add more water into the system, and it involved the basic technique of installing underground tiles to divert water to the nearest waterway. For Parshall, engineered solutions, many of them quite simple, would alleviate the Piedmont's water woes and foster a more productive and lucrative agriculture.[50]

Then why were these solutions not employed more consistently? Parshall complained that much of the answer lay in addressing the outdated business practices and "lax and crude methods" employed by mutual irrigation companies. According to Parshall, rather than addressing inefficiencies in the system, they focused their energies on demanding more water. Parshall opined that these companies lacked the foresight to plan for their water futures and consequently were unwilling to invest the money necessary to upgrade their aging systems. He further complained that mutual irrigation companies were "managed by men with limited business experience" who "lack[ed] aggressiveness and individuality of purpose." Believing that effective management required centralized control over irrigation waters, Parshall argued that as the region's population continued to increase and the demand for water with it, the region's irrigation required specialists with business acumen who could manage water effectively.[51]

The Parshall Flume

Parshall's support of technocratic oversight was energized by a well-known secret exposed years earlier by Elwood Mead, CAC's first irrigation engineer. Mead, who became chief of the Federal Bureau of Reclamation in 1924, observed that mutual irrigation companies colluded with Colorado's water courts to obtain far more water than their members needed, in some case as much as fifty times more than their lands required. This enabled them to profit from farmers with less secure water rights by charging them exorbitant rates for water on demand. This violated the concept of "beneficial use," which was written into Colorado's laws and animated by the principal that water rights accrued to individuals and were limited by the quantity users could put to personal economic benefit. In theory, this protected the idea that water was for the public good, not private speculation. In practice, Mead stated, failure to adhere to the concept of beneficial use was akin to plundering the public treasury. Writing in 1903, Mead argued that scientific expertise and accountability could rein in water scofflaws and result in water equity.[52] Ralph Parshall not only agreed with Mead's assessment but oriented his life's research around chipping away at the inequalities his predecessor had exposed.

Ralph Parshall's greatest contribution to efficient water management came from a device he invented with the help of CAC graduate students. Eventually named after him, it was called the Parshall Flume, and it revolutionized the measuring of water in canals and ditches and, consequently, made water diversion more equitable. As Parshall saw it, the greatest barrier to efficient water management was measurement. He argued that at least 25–30 percent of water was unavailable to more recent settlers on the Piedmont, due to measurement inaccuracies that enabled those with more secure water rights to take far more than their legally allotted share. Parshall further demonstrated that only one-quarter of all water in the South Platte Watershed was measured accurately during the early 1920s. The most common devices at the time used to measure water volume and flow were the weir, generally installed at diversion dams and in canals and ditches, and the rating flume, more commonly placed in streams. Both devices were calibrated to measure the volume of water passing a particular point based on a known flow and depth of water in the channel. This enabled irrigation companies to measure the water passing through their main canal, into their reservoirs, and through the ditches that carried water directly to farmers. Unfortunately, the devices lost accuracy after installation. In numerous writings, speeches, and radio addresses, Parshall harped on how changing conditions caused gross misreadings. Conditions upstream of weirs and flumes altered depth and flow. Gravel and sand might build up in a particular spot, and tree branches and debris clogged waterways, forcing water to pool at greater depths in one location, while causing it to flow swift and shallow in others. All of these hindered accurate gauge readings. The floors of the flumes and weirs were also problematic. Irrigators installed the devices in a level spot on channel floors so that depth could be accurately measured. This presented particular problems when sand and gravel accumulated on the flume. As this occurred, the gauges misread water depth, leading managers to misrepresent the quantity of water in the channel.[53]

In 1921 Parshall and his students claimed to have solved that problem through what they originally called the Improved Venturi Flume (though not named after Parshall until 1929, I will refer to these flumes as Parshall Flumes). Parshall capitalized on the previous work of hydrologists who found a unique relationship between water and flow.

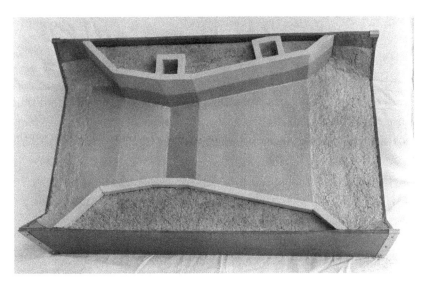

Fig. 20. Model of the Parshall Flume. Irrigation Research Papers, CSU Morgan Library Archives and Special Collections. Used by permission.

Fig. 21. Ralph Parshall checks on one of his flumes. Irrigation Research Papers, CSU Morgan Library Archives and Special Collections. Used by permission.

In streams where water slows and appears to stagnate, deep pools are often formed. At the downstream end of that pool, the flow increases and the stream becomes shallow. Hydrologists identified what they called "critical flow," the point where water transitioned from slow and deep to fast and shallow. They also calculated a depth at which that flow could be measured. Parshall and his students sought to develop a flume that could generate critical flows in irrigation ditches. These flows were ordinarily not possible in irrigation waterways since engineers and farmers dug their ditches with minimal slope, resulting in slow-moving waters that were ideal for delivering water but not for measuring it. To solve this, Parshall placed his flume on the floor of irrigation canals and ditches, across their entire width. Water entered the flume and converged into a narrow section called the throat. In the throat, Parshall introduced a downward slope significant enough to replicate the critical flow found in natural streams. He then installed a water gauge that could measure the volume. Water then emerged out of the throat and eventually back into the channel.[54]

The flume effectively solved the problem of upstream obstacles since critical flow was introduced within the flume. Upstream conditions might impede water but not its measurement. The Parshall Flume also minimized complications from debris within the flume itself. Critical flow generally introduced enough force to clear obstacles that might otherwise accumulate on the flume floor. According to tests made by Parshall and his students, this new innovation had an error factor of less than 3 percent. If Parshall's estimates were correct, the flume could make 20 percent more water available to junior irrigators in the South Platte Watershed by preventing senior appropriators from taking more than their allotted share. By 1927 the Parshall Flume was already being used in Hawaii, Canada, Central and South America, and Africa. Yet most Piedmont irrigation companies did not rush to upgrade their aging systems.[55]

Why not? The answer is found more in the practical politics of water than in its engineering. The measurement inaccuracies generated by older technologies supported the claims of users with secure water rights for more water than they were entitled to. Poor measurement enabled water fraud. Installing new Parshall Flumes would require mutual irrigation companies to collect money from their members for the installation of devices that would likely reduce their allocation of water by the simple act of accurate measurement. Further, when it came to their most important resource, farmers and their irrigation companies did not wish to invest money in new technology, especially when it might curb their water. As Colorado State University irrigation engineer Chris Thornton pointed out, "There is something inherently conservative in ditch companies; they don't want to spend money and they don't want to mess with [their water]." As long as ditches were full, farmers had little interest in reengineering the system that delivered the resource.[56]

The failure to quickly integrate these new flumes into Piedmont agriculture figured heavily into Parshall's crusade for efficiency and in the specifics of his message. From the late 1920s through the 1930s, Parshall was a regular guest on KOA, a Piedmont radio station emphasizing agriculture. Since Parshall grew up farming irrigated lands in Northern Colorado, he understood how to appeal to this target audience. During several interviews, he argued that accurate water measurement would result in better farmers. Pointing out that too much irrigation could be just as detrimental to a crop as too little, he stated that Parshall Flumes would enable farmers to know exactly how much water they were applying to their crops, facilitating better harvests and more wealth. Speaking to junior appropriators in the region, he called attention to their stresses over water scarcity, stating that better measurement would result in more water security since senior appropriators could not take more than their share. He also appealed to the time and money saved through technical improvements, arguing that the flows generated by the Parshall Flume made it effectively self-cleaning, saving the time and money needed to clean debris from ditches. Parshall tugged at notions of fairness and logic, arguing, "It is just as reasonable to have one's water measured as measuring one's crop for sale." Through each of his broadcasts, Parshall played the role of educator, explaining how his flume operated, hoping for converts among those who doubted its engineering or were repelled by their own lack of installation know-how. Parshall had long been a champion of technocratic efficiency. The radio offered him the platform to broadcast his ideas.[57]

To ensure broader usage, Parshall went beyond airwave rhetoric by designing flumes calibrated to meet the needs of individual irrigators, even participating in some of the installation work. From the time that Parshall developed his namesake flume until the 1950s, he worked to improve the device. Motivated by the varied needs of farmers and the demands of industry and municipalities for accurate measurement, Parshall developed flumes with throat widths ranging from three inches to fifty feet that could measure water flowing at rates up to three thousand cubic feet per second. Due to variances in flow, depth, and site conditions, Parshall calibrated flumes to meet a panoply of needs. He often consulted on projects, visiting an installation site and working with users to select the device most suited to needs. In the early 1950s, Parshall wrote his final articles on the installation and use of small Parshall Flumes. The articles, though published in routine and rigorous scientific fashion, represented more than four decades of dedication to water measurement.[58]

Evaluating the success of Parshall and his flume requires differing metrics. As an engineer, Parshall had a great impact. In the long-term, the Parshall Flume was a resounding success. Worldwide, variations on it remain the standard for measuring water in a ditch or canal, and it has been adapted to a variety of industrial uses. Further, Parshall's reputation as a steward of the public good has grown exponentially. Since neither he nor CAC patented the design (though Parshall did patent the measuring gauge inside the flume), it could be reproduced cheaply. However, during his own time and on the Piedmont—where Parshall grew up and spent his career—the Parshall Flume was only slowly adopted, and often not as a result of either his logical and impassioned crusades or his willingness to adapt the flume to meet local needs. In fact, many irrigation companies did not install Parshall Flumes until the 1950s, when crumbling infrastructure and attractive federal loans conspired to push them toward an efficiency that could have been achieved decades earlier.[59]

Ralph Parshall's work and promotion of the flume were not indicative of an overbearing federal bureaucracy, nor did they reflect the triumph of agri-state science over water management in the West. On the contrary, Parshall had to convince farmers and their irrigation companies that efficient water measurement was in their best interests. Even Parshall's laboratory research reflected this dynamic. In order to develop his original flume, Parshall entered into a nonbinding agreement with the Jackson Ditch Company to build an experimental hydrology lab adjacent to the Cache la Poudre River, in the town of Bellvue, that could accommodate a much wider variance of flows than the facilities at CAC. Once the Parshall Flume was developed, Parshall, along with fellow USDA researchers and graduate students at CAC, employed the Bellvue lab to calibrate individual flumes to meet the requirements of irrigation companies and individual farmers. When looking back at Parshall's career, retired Colorado State hydrologist Thomas Trout observed that Parshall often conducted himself more like an agricultural extension agent than a USDA scientist since he logged so many hours adapting technology to farmers' needs.[60] Regardless of his job title or employer, Ralph Parshall's influence on the Piedmont was tethered to his willingness to promote his flume and then tailor it to meet farmers' demands.

When it came to the Piedmont beet sugar industry, there is no question that the intertwined work of agri-state scientists played a prominent role. Yet influence should not be confused with institutional power. On the Piedmont, agri-state sway resulted from diffusion and malleability rather than centralization and hierarchy. To achieve greater relevance, federal and state researchers within the USDA and at land grant colleges learned that they could not be autonomous actors. Impact required that scientists embrace industry prerogatives, respond to company requests, and provide land and expertise to agro-industry. It was a negotiated influence in which agri-state scientists initially had to convince industry of the value of their work by responding directly to its needs and then craft research agendas that mapped onto its goals. Employing beet sugar industry objectives to concoct their research, they developed joint projects, shared findings and workspace, and blurred lines between government agencies and private companies. As researchers chose projects whose findings could be disseminated broadly through the beet sugar industry, they not only collaborated with Great Western Sugar and its agro-industrial partners but became their employees. To a degree, the ubiquity of state-sponsored science on the Piedmont and throughout the West was a function of its invisibility. Nonetheless, it became a permanent fixture.

The relationships established during the first three decades of the twentieth century between state-sponsored science, industry, and water users on the Piedmont would become more entrenched during the wrenching transitions of the Great Depression. As the beet sugar industry and its growers suffered economically, they emphasized reducing production costs, resulting in further attempts to place heavy burdens on families of contract laborers. Stepping into the fray, federal legislators passed legislation that would redirect agri-state research toward mechanizing Piedmont fields. All of this occurred in the midst of the region's most notable twentieth-century drought, placing Piedmont irrigation engineers such as Ralph Parshall at the center of efforts to divert water from wetter watersheds. Taken together, these events began to erode the Piedmont's agroecological foundations.

5 The Economics of Mechanization and Watershed Engineering

> The more complex and powerful the system of farm produc-
> tion, the more sensitive and strict must be the moral con-
> sciousness behind it and the more elaborate and expensive
> the system of public control overseeing it.
>
> —DONALD WORSTER

From the late 1920s through the 1930s, an Irish Catholic reformer
named Thomas Mahony posed a threat to the labor model employed
by Great Western Sugar and its growers. A businessman who moved
to Longmont in 1920 with only a fifth-grade education, Mahony was
appalled by the contract labor system employed by the sugar company and its growers. Consequently, he and several congregants from
Piedmont Catholic parishes formed a fraternal organization called the
Mexican Welfare Committee of the Colorado Knights of Columbus.
Their stated purpose was to "improve the religious, social, and economic conditions affecting our Spanish-speaking fellow Catholics,"
to address the living and working conditions of migrant workers,
and to promote "fair dealing with them." The Committee offered
economic aid to laborers, fielded worker complaints, and compiled a
host of data on living and working conditions, which it then used to
expose exploitative practices. Mahony chaired the Mexican Welfare
Committee and was its most outspoken activist.[1]

In May of 1929, Mahony's efforts began to yield substantial fruit
when the committee published a speech he delivered to the Catholic

Conference on Industrial Problems in Denver. Titled "Wages of the Unskilled Workers in Colorado," it employed labor data and evidence gathered by the committee for nearly a decade. Mahony calculated that a family of five or six needed an annual income of $1,800 to live "at a minimum level of health and decency." Yet, the wages of entire families working in the beet fields averaged $600. Mahony opined that the "skill, intelligence, and conscientious work" of hand laborers was largely responsible for healthy crops, yet the financial rewards were disproportionately reaped by growers and Great Western. Transitioning into housing, Mahony noted the gulf between contract terms and lived reality. Contracts stated that "habitable houses and suitable water near at hand for drinking and domestic purposes" should be provided by the grower and verified by Great Western. Still, the majority of families continued to live in "one or two-room shacks" that were "poorly lighted and ventilated . . . not properly screened," and often lacked sanitary water nearby. Finally, Mahony connected those marginal living and working conditions with children missing school—despite Colorado's compulsory attendance laws—to work in the fields, holding Great Western, growers, and public officials responsible for their truancy.[2] After Mahony delivered the speech, the Knights of Columbus printed it and distributed thousands of copies. Many of them were delivered into the hands of leaders within the American Catholic Church. From there, the speech found its way into the hands of secular magazines such as *New Republic* and *The Nation* that published articles and editorials drawing national attention to conditions in Piedmont beet fields. Consequently, Mahony became a Piedmont point of contact for reformers, academics, and writers looking to expose industry injustices.[3]

For C. V. Maddux, Great Western Sugar's chief labor recruiter, Mahony was more than a nuisance. Since hand laborers performed two-thirds of the work required to move beets from seed to factory and their wages represented over half of the money paid out by growers to cultivate beets, Maddux could ill afford to cede the labor narrative to this local reformer. In speeches and writings Maddux countered that Mahony's wage claims were inaccurate and misleading.

He stated that families earned $850 annually in addition to receiving free housing and transportation. Disputing the idea that these wages were insufficient, Maddux argued that their pay reflected work performed for only half of the year. Further, he blamed laborers for their struggles. Employing common 1920s racial stereotypes, Maddux claimed that Mexicans were incapable of dealing with periods of distress without outside aid and that they needed to learn to "save for a rainy day." Maddux then explained that Great Western pressured growers to improve worker housing, while providing land for the construction of worker-owned homes, or *colonias*, in most beet districts. He argued that company attempts to improve conditions resulted in more laborers and their families remaining on the Piedmont year-round, stating that their numbers had increased fourfold from 1922 to 1928. In Maddux's telling, dignified work at a fair wage drew beetworkers to the Piedmont.[4]

On the eve of the Great Depression, the tension between Mahony and Maddux exposed the unstable foundation of Piedmont agriculture and foretold cracks in its future. Since the beet sugar industry's establishment in 1901, interdependency characterized the energies that held it together. Great Western's growers added sugar beets to preexisting crop rotations, enhancing an already productive agriculture with a vegetable whose byproducts fed livestock and replenished soils, adding to the region's wealth. By 1930 the critical mass of humans energizing the system were Mexican nationals and Mexican Americans, the majority of whom worked as families and migrated seasonally to the Piedmont. Growers and Great Western alike claimed that, without a critical mass of these low-wage laborers, the Piedmont's economy and system of sustainable farming would collapse. So, when reformers such as Thomas Mahony pulled back the curtain on injustices in the contract labor system, they also threatened to destabilize the energies that held together Piedmont agroecology.

The economic collapse of the 1930s provided the impetus for mechanizing Piedmont labor. As sugar prices and consumption dropped for the first sustained period in the twentieth century and reformers exposed the plight of beetworkers, the federal government stepped in

U.S. Sugar Consumption by Source, 1925-1934

| Year | Consumption (tons) | United States Contributions | | Cane Sugar from U.S. Colonies | Other Contributions |
		U.S. Beet Sugar	U.S. Cane Sugar		Imported Sugar
1925	6,603,000	16.13%	2.27%	28.10%	53.50%
1926	6,796,500	15.39%	1.24%	24.62%	58.75%
1927	6,348,000	14.73%	0.73%	29.43%	55.11%
1928	6,642,500	18.72%	2.09%	31.61%	47.59%
1929	6,964,000	14.74%	2.71%	30.40%	52.14%
1930	6,710,500	16.99%	2.94%	35.73%	44.34%
1931	6,561,500	20.44%	3.14%	38.55%	37.87%
1932	6,248,500	21.10%	2.56%	47.71%	28.63%
1933	6,316,000	21.63%	4.99%	47.91%	25.47%
1934	6,476,000	24.03%	4.03%	42.31%	29.63%

Fig. 22. U.S. sugar consumption by source, 1925–34. Graphic created by Philip Riggs and adapted from Ray Ballinger, *A History of Sugar Marketing through 1974*.

to stabilize the industry through a combination of subsidies, quotas, wage guarantees, and child labor reforms.[5] In response, Great Western and its growers repurposed the economic crises and government intervention into a clarion call to mechanize its workforce, enlisting agri-state scientists to subsidize industry agendas. Land grant colleges in beet-growing states collaborated with several USDA agencies to produce cultivators, harvesting machinery, and seed varieties aimed at eliminating human energies in the beet fields. Paralleling these developments, USDA and Bureau of Reclamation engineers collaborated with Great Western and its growers to provide the economic rationale and technical skills needed to invigorate Piedmont fields with water from across the Rocky Mountains. Taken together, these efforts represented a substantial shift in the acquisition and application of energies for managing Piedmont agroecology, opening doors for entirely remaking it in the future.

"Hunger Pressure"

The tariff provided the economic foundation for the Piedmont beet sugar industry and sustained it through the first three decades of its existence. So, with wholesale prices for sugar dropping to their lowest levels in the twentieth century, beet sugar industrialists fell back on existing strategy. In 1930 beet sugar industry leaders clamored for and received a tariff increase of almost 12 percent on imported sugar through the Hawley Smoot Tariff. With the costs of other farm commodities in free fall, the wall of protection against foreign sugar— primarily from Cuba—convinced Piedmont growers that planting large beet crops was the lesser of many evils. As prices for farm commodities nationwide dropped an average of 60 percent from 1929 to 1933, prices for sugar beets fell a modest 30 percent, due primarily to restrictions on Cuban imports.[6]

Though the tariff cushioned the Depressions' economic blow, Great Western and its growers sought further protection from financial losses by transferring a disproportionate share of its losses onto laborers. In the spring of 1930, contract workers came to the fields still reeling from the disastrous 1929 harvest, when the sugar company delayed the harvest and early frosts forced them to forfeit wages to supplemental laborers who were hired by growers to pull and top beets. Laborers arrived in 1930 to find that Great Western and its growers had radically altered their terms of employment, crafting contracts that paid hand laborers a percentage of earnings, instead of a flat rate per acre. Under this system, growers received their beet checks from Great Western, and then laborers were given a percentage of that check—typically 20–25 percent. Growers also resorted to paying by the job. So, instead of being hired on for an entire season, they hired some hand laborers only for thinning, cultivating, or harvesting. Before the Depression, such practices would have decimated the labor pool. Desperate for work and lacking advance knowledge of the contract terms, laborers had little choice but to accept them.[7]

Laborers felt the impacts immediately. At the conclusion of the first season under the new terms, a fully employed family of contract laborers averaged less than $300 annually. Over the next two years, conditions deteriorated further. The *Colorado Labor Advocate* found that wages in 1933 had dropped to $120–$130 annually, when taking into account that many laborers were not fully employed. Using the term "industrial slavery" to describe the conditions in the beet fields, its author contrasted this remuneration with the 7 percent dividends Great Western still paid on its common stock that year.[8] Offering another perspective to view the disproportionate wage drop, Thomas Mahony compared the tariff benefits received by Great Western with the wages paid out to workers. He stated, "The tariff on last year's production of sugar is more by several million dollars than the total amount paid the Colorado farmers for beets, the 28,000 field hands for their labor and the 5,500 or so factory workers . . . all taken together."[9] Suggesting that this was the intention from the start, Reed Smoot of Utah, one of the authors of the new tariff and a senator from another state with a sizable beet sugar industry, predicted in tariff hearings that the beet sugar industry "will not have to pay a dollar more for Mexican labor than in the past." The high tariff offered the beet sugar industry some protection from the worst storms of those early Depression years, while laborers bore the full brunt of its economic fury.[10]

As desperation gripped the Piedmont, beet workers depended on public assistance more than ever before, at a time when social services were strained to their breaking point. This was not entirely new. Most Piedmont counties maintained a small "poor fund" whose monies could be used to provide a pittance for families in need. Larger cities, such as Denver and Pueblo, where sizable numbers of Mexicans lived during the off-season, provided more public assistance per capita than Piedmont towns. Various private agencies, such as Colorado Catholic Charities, also provided health clinics, self-improvement classes, food, and monetary aid.[11] The scale of the need, however, overwhelmed local governments and private charities, shifting the burden to federal relief agencies.[12]

Several examples demonstrate that shift toward federal aid. In 1933 the Federal Children's Bureau compiled several case studies of families in the beet fields who wintered in Denver and Pueblo, most of whom worked for Great Western Sugar. Frank M. was one of Great Western's favored workers. During the 1920s the company offered him the chance to buy a small lot and build an inexpensive house within a colonia. He lived in a two-room house he built for his family of seven. Owing to falling wages, he made no house payments after 1931. He was able to do some side work by cultivating other crops, but he began to receive federal aid in 1932 for four months out of the year. In 1933 he had a percentage contract that did not pay him until the end of the season, requiring him to take in federal relief at the rate of twelve dollars per month. One of his peers, Manual G., also with a family of seven, was fortunate to live rent free in a house owned by Adams County. In 1931 he worked ten acres at ten dollars per acre and received supplemental work picking beans and tomatoes, but his money ran out in January, requiring him to apply for federal aid. He could not get a beet contract in 1932, so he picked beans again, relying on partial federal aid. He received a thirty-acre contract in 1933, and subsequently bought a truck for $200, hoping to gain employment as a hauler in the off-season.[13]

Stories of struggles like these informed the increasingly confrontational tactics of sugar beet labor unions in the early 1930s.[14] In 1932 beet workers on the Piedmont formed the United Front Committee of Agricultural Workers. The United Front was a subsidiary of the communist-led Trade Union League (TUUL). According to historian Dennis Nodín Valdés, TUUL's presence may have originated when several Colorado beet workers attended TUUL's first Ohio meeting in 1929. In May 1932, after most migrant laborers arrived on the Piedmont, and before the start of thinning season, the United Front called for a general strike by all beet workers on the Piedmont, to protest wage reductions by growers. An estimated eighteen thousand workers refused to take up their hoes and shovels that spring in response.[15]

Though large, the general strike was quickly quelled as growers and Great Western used several tools to break union resolve. The foremost among them consisted of what Thomas Mahony referred to as "hunger pressure." Since beet workers relied on public aid to survive, Great Western and its growers employed their influence with local relief boards to remove striking workers from the rolls. In Weld County, sheriffs patrolled farms to force workers into the fields, jailing or deporting those who refused. In fact, between 1930 and 1935, local sheriffs, relief agencies, and the Immigration and Naturalization Service colluded to deport twenty thousand Mexicans living in Colorado.[16] Since the beet sugar industry employed more Mexicans than any other industry in the state, it is likely that at least half of those deported were Piedmont beetworkers. The strike also failed due to overreach. Communist labor advocates from TUUL and International Labor Defense (ILD) had pledged financial and legal support for the strikers that largely did not surface.[17] Finally, with a Depression-induced labor surplus, growers could cobble together enough workers to carry on cultivation as union resolve crumbled. So, the 1932 strike collapsed, and the union quickly disintegrated.[18]

Though it ultimately failed, the strike was just one more example of the essential position contract laborers occupied in Great Western's attempts to harness Piedmont energies for sugar production. The company's collusion with immigration authorities and local law enforcement and relief agencies demonstrates its desperation to retain a critical mass of cheap labor even as its strongarm tactics would likely erode the company's pool of seasonal laborers. After 1932, migrant workers increasingly thought twice before crossing the southern border to work for Great Western, often seeking other employment when they did. Company practices also provided more fuel to the fire of labor reformers such as Thomas Mahony who, by the early 1930s, was in consistent communication with the Mexican Consulate, relaying messages about conditions in the beet fields. Spanish language magazines such as *La Prensa* exposed Great Western's practices as well. Consequently, the opportunistic messages delivered by C. V. Maddux and his team of labor recruiters competed with narratives that evoked fear and reticence.[19] For long-time beetworkers such as Manual G. and Frank M. and those like them, insecurity in the fields encouraged them to look elsewhere for temporary employment that could turn into permanent separation from Great Western. The company's business model, like that of its industrial peers, emphasized harnessing the energies found in human bodies, along with those found in soil, air, and water, to produce a commodity. The Depression placed that model into chaos and reformers exposed how the human portion of it was built on exploitation.

The Sugar Act

As the fictional narrative about humane labor in the beet fields eroded, so did the economic stability of the tariff, pushing the federal government to intervene. While beet sugar production in the United States remained high, the tariff could not stop wholesale prices from plummeting. U.S. colonies such as the Philippines, Hawaii, and the Puerto Rico increased their production of raw cane sugar since it entered the United States duty-free. By contrast, Cuban sugar imports dropped by half from 1929 to 1933 as the Hawley Smoot Tariff decimated Cuba's most important industry, bringing it to the verge of wholesale bankruptcy. To address the instability, the U.S. Tariff Commission intervened in April 1933, recommending that the United States reduce its duties on Cuban sugar and instead adopt a quota system that would regulate the amount of sugar from the domestic, colonial, and foreign sources. This synchronized with the farm policy being developed by the Roosevelt administration and a democratic Congress. In March 1933, Congress passed the Agricultural Adjustment Act (AAA), which created quotas for seven basic farm commodities. Seeking to bring production in line with consumption, the federal government issued farmers checks based on acres they took out of production in efforts to reduce supply.[20] Sugar presented a unique crisis since it required setting quotas that balanced domestic and imported sugar. In June 1933, Secretary of Agriculture Henry Wallace formed the Sugar Stabilization Committee, made up primarily of leaders within the domestic beet and cane sugar industry, to draft an agreement that would strike a balance between sugar supply and consumption.[21]

Though focused on the prerogatives of sugar processors, the Stabilization Committee also took testimony from growers and other interested parties. Even so, the appearance of beetworker Leo Rodriguez offered a striking contrast. A U.S. citizen who worked as a hand laborer in Piedmont beet fields since 1903, Rodriguez went to Washington to tell the history he and his family shared of working in the beet fields. Existing evidence suggests that Rodriguez maintained a positive relationship with Great Western for most of his career. In 1926 he penned a poem for a local newspaper that showed his appreciation and may have been used as a company recruiting tool. Peppering common Spanish words in simple rhyming verse, Rodriguez extolled the company's virtues with phrases such as these:

> They [Great Western] are always on the square
>> When it comes to your Dinero;
> They pay it right in hand
>> So that is pretty Bueno.[22]

Rodriguez continued to work on the Piedmont during the Depression, living in Pueblo with his family of nine during the winter and laboring in one of Pueblo's steel mills. Work at the mill dried up even as beet wages plummeted, and Rodriguez went on relief. He stated that each of his neighbors were beetworkers, and all but one of them received aid. For a month, his family of nine received "one-hundred pounds of flour, a nine-dollar order of groceries, and a half-ton of coal." Rodriguez also opined that fourteen of his peers were cheated out of their wages by growers in 1932. He told the committee that when he did labor in the beet fields, he earned an average of four cents per hour, requiring him to work each of his able-bodied children. Rodriguez asked the committee to reform the system so that he could "send [his] children to school."[23]

Rodriguez's history of hard work and devotion to the sugar industry and his U.S. citizenship are likely reasons why the Federal Children's Bureau, the National Child Labor Committee, and the Catholic Conference on Industrial Problems took him to Washington. It was not plausible to paint Rodriguez with the kind of racial stereotypes that had been used to marginalize Mexican beetworkers for decades. Rodriguez was neither a peon nor a radical. Here was a devoted laborer with a strong resume of loyalty to Great Western Sugar and its growers who, nonetheless, put his entire family to work in the beet fields as part of the industry's contract labor system. If anyone deserved a fair shake from the beet sugar industry, it was Rodriguez. Yet, there he was, giving testimony, dressed in fieldwork attire, and scraping by with the help of federal relief. It was a clear message to the Sugar Stabilization Committee that honest attempts to reform the beet sugar industry required improving the livelihoods of its laborers. Rodriguez continued to advocate in early 1934, when he supported a petition from beetworkers to Secretary Wallace. In addition to corroborating Rodriguez's earlier testimony on wages and relief, the petition stated that laborers and their entire families were being pushed into a state of "permanent peonage," and that growers continued to take advantage of the naiveté and desperation of Mexican workers to force them to sign oppressive contracts. Finally, they complained that the local relief agencies responsible for administering federal aid colluded with Great Western and its growers to force Mexicans into the fields whenever their labor was needed.[24]

Despite some obstruction by beet sugar processors, the final version of the Jones-Costigan Amendment to the AAA—also called the Sugar Act—passed in 1934. Though it contained some provisions destined to improve conditions in the beet fields, its initial administration demonstrated a lack of concern for the wages of workers. It added sugar to the list of agricultural commodities in the AAA and set a voluntary production quota of 1.55 million tons of beet sugar that could be refined in the United States, a quota that the entire industry agreed on. The remainder of the nation's estimated 6.5 million tons of sugar consumption would come primarily from Cuba, the Philippines, Hawaii, and Puerto Rico. Based on sugar beet acreages in 1934, growers were scheduled to receive $10 million in federal payouts for reducing their production. To receive payouts, growers agreed that no children under the age of fourteen would be employed, and that those between fourteen and sixteen would be required to obtain work permits and could not labor more than eight hours per day. As part of the Sugar Act, the Secretary of Agriculture set annual wages and timetables of payment for laborers in the beet fields, requiring growers to certify that they had paid their workers before receiving their federal refunds. Growers were also required to provide "a habitable house, suitable water near at hand for drinking and domestic purposes, a suitable garden plot . . . and transportation . . . to the farm prior to the beginning of the hand labor operations and from the farm . . . upon completion of the work contracted."[25]

The crafting and implementation of the Sugar Act emphasized the marginality of sugar beet laborers. When the bill went into effect in May 1934, beet growers were immediately apprised of their allotted acreage, knew how much they would be paid per ton of beets, and could count on a check at the end of the season for unplanted acres. Laborers, on the other hand, possessed no such security. As late as the end of August, Wallace had not published wage rates, so most laborers were not paid a dime. In lieu of this, many growers attempted to foist the same percentage contract on workers used in previous years, some claiming that this was in fact the contract that had been approved by the Secretary of Agriculture. With reduced acreage and no guaranteed contract, some beet workers hoped that more federal aid would be extended to them.[26]

It was not. As much as 91 percent of federal aid administered in Colorado during any given Depression year after 1932 came from the Works Progress Administration (WPA). The WPA made aid determinations based on labor statistics from states. Counties, however, administered the aid. On the Piedmont, Great Western's representatives and growers dominated the county relief boards. They used that influence to remove beet laborers from county relief rolls in May and June, at thinning season, and again in August as preparations for the harvest began, employing a tactic Thomas Mahony referred to as "hunger pressure methods."[27] Though Great Western provided a list of laborers to WPA boards, when it was time to cull its lists, it was common practice in Piedmont counties to simply remove everyone with a Hispanic surname from relief rolls. Great Western often took advantage of the sudden influx of labor to place beetworkers into districts in Nebraska and Wyoming, where labor shortages existed.[28] In Weld County, beetworkers who sought to return to federal relief rolls after the harvest had to submit a special "Application for Federal Relief" that required a signature from a Great Western fieldman before they could receive aid. It asked questions regarding whether they had conserved food, whether they had gardened successfully, and what type of clothing they possessed. This implied that Mexican beet workers were not thrifty with their resources and that their employer, Great Western, could require frugality as a precondition for federal aid.[29]

In the midst of these heavy-handed and racist tactics by Great Western and its growers to control labor, it is easy miss that these were stopgap measures that thrived in a year of transition. Once the Sugar Act was fully understood by laborers, their wages increased as growers were compelled to pay them at least the wage published by the secretary of agriculture prior to receiving a subsidy check from the federal government. Further, while child labor remained common, in violation of the Sugar Act, fewer workers sent their entire families into the fields and neither growers nor Great Western could so easily plead ignorance when violating the law.[30] Federal law was not the only factor that reduced exploitation. The work of social reformers such as Thomas Mahony highlighted the living and working conditions of beet laborers for a national audience. Meanwhile, labor agitation and the testimony of people like Leo Rodriguez emphasized that the sugar beet industry could not so easily contrast labor abuses in the tropical cane sugar industry with conditions in the Piedmont sugar beet fields. The human muscle necessary to cultivate beets remained exploited. But beet laborers were emerging from the shadows of obscurity and commanding a larger share of the sugar beet pie than they had in the past. As Great Western's ability to shape the labor narrative eroded, the company sought to squeeze more sugar out of each hour of work and, ultimately, to eliminate laborers altogether.

Mechanizing the Piedmont

In its mandate to regulate sugar production and wages, the Sugar Act had a profound impact on the science and technology of sugar beet cultivation. To carry out sugar quotas, the secretary of agriculture assigned refiners such as Great Western a limited number of acres for beet cultivation based on estimates of how the United States would meet its annual sugar quotas. While quotas regulated land use, they did not cap the quantity of sugar growers could wring out of each plot of land. Without an incentive to expand onto new lands, Great Western and its growers prioritized per acre tonnage and sucrose content even more than they had previously. However, even if more sugar could be squeezed out of each acre, contract workers still presented the greatest obstacle to bolstering profit under the new law as they still represented more than 50 percent of labor costs. Since the Sugar Act essentially created minimum wages for workers and cracked down on child labor, growers were hard-pressed to reduce the share of income paid out to laborers without replacing them with machines. So, once again, sugar companies enlisted the aid of agri-state scientists, this time to accelerate research aimed at increasing per-acre sugar content and mechanizing cultivation.[31]

Prior to 1934, efforts by the beet sugar industry and agri-state scientists to mechanize operations proceeded on a piecemeal basis, reducing the number of labor hours per acre by 20 percent as compared to 1909. Hand-operated cultivators that cleared weeds between rows were in use since the Piedmont sugar beet industry's inception in 1901. By 1934, four-row cultivators, pulled by horses, were the standard. During the 1920s, growers used horse-drawn beet pullers and mechanical loaders for harvest operations. These reduced the labor necessary to extract beets from the soil and load them into wagons. By 1930, growers were trading in their wagons and replacing them with trucks for hauling beets. While some growers used machines for blocking in the spring and cultivating rows of beets during the summer, they had not gained wide usage on the Piedmont due to cost and the fact that they compared poorly with the precision of hand labor. As late as 1936, few growers employed tractors in their operations. Though tractors were commonplace in the larger, more heavily capitalized sugar beet operations in California by the mid-1930s, one 1936 study showed that twenty-five of twenty-eight Piedmont farmers employed only horses.[32] Where seeds were concerned, the work of agri-state scientists and Great Western Sugar supplied Piedmont growers with varieties adapted to the local climate. However, seeds bred to resist disease were not yet available, and there was no promise of monogerm seeds on the horizon. While fertilizers boosted beet yields elsewhere—most notably in Michigan and the Midwest—effective crop rotations, livestock manures, and nitrogen-fixing legumes nullified the need to add fertilizer on the Piedmont. As for pesticides, Great Western supplied its growers with Paris Green and sprayers during periodic infestations. All factors considered, mechanizing labor represented the most critical frontier for reducing industry dependence on imported labor.[33]

After Congress passed the Sugar Act, collaboration between the beet sugar industry and agri-state scientists intensified. In 1931 the USDA's Division of Farm Power and Machinery began collaborating with engineers at the University of California and CAC to coordinate mechanization efforts. Researchers developed and tested machinery, distributing findings throughout the beet sugar industry. While their work yielded modest advances in its first few years, the Sugar Act incentivized efforts. Industry leaders such as Great Western Sugar president Frank Kemp and trade groups such as the U.S. Sugar Beet Association lent financial support to agri-state efforts. This resulted in a flurry of efforts to build harvesters that could pull, top, and load beets in a single operation. During a sixteen-month period from 1937 to 1938, researchers tested thirty-two recently invented or proposed harvesters, several of which were developed by the sugar companies themselves. Even as they scrutinized new machinery, Great Western's J. B. Powers found a geometric relationship between beet diameter and crown thickness that enabled uniform machine topping of beets, thus eliminating waste in the process. According to researchers, once commercially available, the new machines would immediately reduce harvesting labor by 30–40 percent. At the same time, Great Western Sugar and Crystal Sugar collaborated with plant scientists at the USDA to develop two new seed strains that resisted leaf spot—the most common disease that infected Colorado's beets.[34]

One way to understand the quickening pace of technological change and collaboration in sugar beet fields is through the intertwined careers and research agendas of its prominent researchers. In 1930 Anton Skuderna—former research director with American Sugar and collaborator on agrochemical research in the Arkansas Valley— began working as principal agronomist for the USDA's Division of Sugar Plant Investigations (SPI). Working out of CAC's experiment station in Fort Collins, Skuderna used his position to develop research in cooperation with Great Western Sugar.[35] To aid in that endeavor, Skuderna hired H. E. Brewbaker, a plant geneticist, as an associate agronomist in 1930. Hoping to employ Brewbaker's talents to greatest effect, Great Western enlisted him to write articles aimed at growers in its publication, *Through the Leaves*. During the early 1930s, Brewbaker was the magazine's most consistent contributor, offering abridged versions of his research, and focusing especially on the value of embracing industry innovation. For example, in an article promoting greater adoption of machinery in the fields, Brewbaker spoke to

Fig. 23. Segmenting multigerm sugar beet seeds, ca. 1941–44. Farm Security Administration, Office of War Information photograph collection, Library of Congress, Prints and Photographs Division.

growers in the same language that Great Western employed, declaring that mechanization would help the grower "produce the largest amount of sugar per acre at the lowest cost per unit of production."[36] In Brewbaker's writings, the aims of state-sponsored science mapped onto those of industrial agriculture. Great Western agreed, offering Brewbaker the job of research director in 1937. He accepted.[37]

Researchers such as Brewbaker and Skuderna were able to transition seamlessly back and forth between agri-state and industry science because their projects were collaborative and shared the same goals. Brewbaker immediately demonstrated this reality when Great Western hired him. In 1937 he and Skuderna gathered a diverse group of sugar beet scientists for the inaugural meeting of the American Society of Sugar Beet Technologists (ASSBT). Meeting in Fort Collins, the organization included representatives from agricultural colleges, USDA scientists, representatives from three seed companies, and thirteen beet sugar refiners. Emphasizing seed breeding and mechanization, their purpose was to "foster all phases of sugar beet and beet sugar research, and to act as a clearinghouse for the exchange of ideas resulting from such work." During its first decade, ASSBT grew to include most of the major state and industry figures researching beets.[38]

Out of that collaboration, a critical breakthrough occurred. In 1941 Roy Bainer, a USDA researcher conducting seed trials at University of California, Davis and a regular contributor to ASSBT meetings and reports, developed a machine capable of shearing sugar beet seed germs into individual seeds. Segmented seeds promised to revolutionize the industry. The single seeds enabled seed planting and germination at uniform distances, significantly reducing the need for hand blocking and thinning.[39] The industry immediately jumped aboard the seed shearing train. In 1942 Great Western built four facilities to segment beet seeds, and Vice President D. J. Roach stated that the company would have an unlimited supply of segmented seeds in 1943, predicting that half of its grower acreage would be planted with the new seed. That year, Bainer claimed that sheared seed would save three million hours of labor across the industry.[40] Segmented seed spawned a host of complementary machines, including planters, thinners, and cultivators. The transformation was especially welcomed by industry since World War II had thinned the ranks of available labor.[41] The labor shortage and the possibilities of segmented seed propelled mechanization efforts, as USDA and private researchers developed machines that could loosen, dig, and pile beets in one operation. At war's end, the technological infrastructure to mechanize the entire industry was in place.[42]

Despite its value, seed segmentation proved to be an imperfect transition to mechanization. Shearing resulted in damage to the seed germ and, even though the process was supposed to produce individual sugar beet seeds, 20 percent of all segmented seed germinated in pairs, requiring hand thinning. Moreover, the misshapen seeds that resulted were difficult to calibrate with existing seed drills. Ironically, this push for mechanized uniformity in the industry highlighted the skill and value of a shrinking labor force. In a 1946 report by the USDA-sponsored Sugar Advisory Committee, beet sugar industry leaders confessed that hand labor still resulted in higher yields than machines and that the best they could hope for was mechanization that minimized waste as compared to the precision work of hand labor. While the report went on to advocate for more public funding and collaboration between agri-state science and industry, it revealed that, when it came to wringing each ounce of sugar from every beet, human muscle could not be bested by machine torque. Industry goals, however, revolved around lowering production costs through mechanizing operations. In that regard, a little waste and imprecision were acceptable tradeoffs for eliminating human labor.[43]

In 1948 science delivered the breakthrough that eluded the beet sugar industry for nearly a half century. Three years earlier, the USDA and beet sugar industry representatives coordinated with European scientists to bring V. F. Savitsky, a leading Russian expert on monogerm sugar beets, to the United States. Savitsky then spent three years investigating existing domestic seed stock before honing in on two beet varieties that commonly produced single seeds. Savitsky and his colleagues then crossed the new monogerm seeds with multigerm varieties that had been bred for disease resistance and climate variations. Writing in 1950, H. E. Brewbaker proclaimed the work a success. The longtime USDA scientist and research director for Great Western then began breeding disease-resistant monogerm seeds at the company experiment station in Longmont. That year he predicted that the new seed would be commercially available to every Great Western grower within five to ten years. Brewbaker's prognostications proved accurate. In 1957, F. V. Owen, the principal sugar beet geneticist for the USDA, announced that monogerm seed had finally become a commercial reality.[44]

Monogerm seeds, alongside other developments, quickened the pace of mechanization and reduced dependence on hand labor. In 1950 the combined efforts of agri-state and industry scientists yielded harvest machinery that removed all need for hand labor in that phase of cultivation. Eight years later, once growers and the sugar company purchased a critical mass of the machines, 98 percent of Piedmont beets were harvested mechanically. In addition, Piedmont growers replaced their horses with tractors since fossil-fuel power could be applied in multiple phases of cultivation. Chemicals also thinned the ranks of human labor. In the immediate postwar era, CAC became the site of field trials where agri-state and industry scientists sprayed a bevy of newly developed organic chemicals on sugar beet crops in search of compounds capable of killing weeds without harming desired crops. By the early 1950s, several were in commercial use.[45] Monogerm seeds brought down the final barrier to mechanization. Since these varieties were nearly uniform in size and shape, they paired well with machine planters. More importantly, monogerm seeds germinated as single plants at uniform distances, enabling cultivator blades to cut down most weeds, leaving healthy sugar beet plants in place. Within ten years of their introduction in 1957, monogerm seeds completely replaced their multigerm predecessors. The long-term impact of mechanization in the beet fields was dramatic. In 1920 Piedmont sugar beets required between 80 and 135 hours of labor to produce a ton of sugar. By 1948, that figure dropped by over half. In 1964, 2.7 hours of field labor netted enough beets to refine a ton of sugar.[46]

Ralph Parshall and the Colorado–Big Thompson Project

In the same year that the Sugar Act incentivized the mechanization of Piedmont sugar beet fields, nature stimulated another kind of technology with the potential to reshape the regional landscape with the aid of agri-state engineers. Lower than average snowpack in the Rocky Mountains during the early 1930s contributed to water insecurity, with 1934 shaping up to be the worst year of drought.[47] The previous year, as water scarcity mounted, Piedmont farmers, politicians, promoters, and Great Western Sugar resurrected a transmountain diversion project that eclipsed all previous efforts in scale and scope. Eventually named the Colorado–Big Thompson Project (C-BT), it would tap into the Colorado River's abundant headwaters across the Continental Divide, redirecting them through a tunnel running underneath Rocky Mountain National Park and into a host of rivers, canals, and reservoirs before flowing into Piedmont irrigation ditches.[48] Based on preliminary estimates, advocates projected that 285,000 acre-feet of water could be added annually to the Piedmont's water supply through the C-BT. From the start, supporters conceived the project as a Bureau of Reclamation undertaking. Piedmont agricultural interests hoped federal largesse and engineering expertise could be enlisted once again to support the region's agenda.

In truth, trans-mountain diversions were not novel in the 1930s. The 1882 Colorado Supreme Court decision *Coffin v. Left Hand Ditch Co.* established the legality of such projects. The largest of these, the Grand River Ditch, transported water in an unlined ditch and wooden flumes to Fort Collins through an area that would eventually be added to Rocky Mountain National Park. By the early 1930s, there were several trans-mountain projects that reversed the flow of waters from the wetter west slope of the Rockies to the drier east slope. In all, these projects accounted for more than thirty thousand acre-feet of water, enough to provide sufficient irrigation to grow twelve thousand acres of beets in an average year, or 7.4 percent of the total number of acres in beets in 1936.[49] This was more than a drop in the bucket, but not enough to meet the growing demands of Piedmont farmers and industry.[50]

If successful, the C-BT set would set multiple precedents. Prior to the proposed project, Reclamation had not constructed any projects on the Colorado River Watershed in its upper basin states of Colorado, Wyoming, Utah, and New Mexico, let alone divert water from the river into another basin entirely. Further, engineering water for the Piedmont violated two of Reclamation's core principles. Framers of the 1902 Newlands Act, which created the Bureau of Reclamation, stated that its projects would be for the express purpose of opening new lands for agricultural development. Farmers who moved onto lands blessed with the Reclamation water would only be allowed to access enough liquid gold to water 160 acres. By contrast, the Piedmont was already a well-developed, irrigated agricultural region and, though the majority of its farmers cultivated fewer than 160 acres, its history of water appropriation suggested problems ahead should Reclamation try to enforce its acreage limit. In fact, the Bureau's history to that point demonstrated that it was more committed to building dams and harnessing nature's water wealth for private profit than to any philosophical goals. In the case of the C-BT, however, violating Reclamation's stated aims was baked into the project.[51]

Five separate Piedmont counties and cities as well as Great Western Sugar lined up behind the C-BT. In 1934, Reclamation agreed to conduct engineering studies in advance of a project proposal. The Public Works Administration then provided $150,000 to complete it. Over the next three years, support for the C-BT emerged from a variety of sources, including all but one member of Colorado's Congressional delegation, editors of every Piedmont newspaper, a majority of local elected officials, mutual irrigation companies and their farmers/ stockholders, and academics such as Ralph Parshall.[52]

While a bevy of agri-state engineers, politicians, and administrators played critical roles in the C-BT, Parshall was more clearly situated at their intersection than anyone else. Since the 1910s, he had focused his engineering career on measuring and distributing water to the greatest number of farmers in the most efficient fashion. By the mid-1930s, Parshall's résumé expanded beyond his namesake flume to include self-cleaning devices for removing debris from irrigation canals and coordinating snow surveys that predicted water available to farmers in advance of crop selection. Though Ralph Parshall never wavered from his emphasis on efficiency, he also agreed with irrigators who argued that farming the semiarid Piedmont demanded more water. During the Depression, Parshall recognized that supporting the C-BT would have greater impact than crusading for flumes and fixing leaky reservoirs. So, when the USDA's Bureau of Agricultural Economics and CAC commissioned a report on the economics of the C-BT, Parshall was the clear choice.

Published in January 1937, the report is notable for Parshall's deft use of economic and water statistics to characterize Piedmont farmers, place the region in the national imagination, and argue that the C-BT was a wise financial decision. In the report, Parshall, aware that a massive Reclamation project might be viewed as a government handout to wealthy farmers during the Depression, characterized Piedmont farmers as "hardy, self-reliant American farmers and townspeople" who needed additional water to "stabilize the present economic achievement and make secure the possibilities of future progress." In a nod to popular New Deal programs, Parshall stated that the guarantee of sufficient water would be like "social security" for existing farmers, enabling them to gain the same comfort in their later years that was newly available to working class Americans. Hoping to demonstrate that the C-BT was a difference maker, Parshall argued that its greatest value was that its water would be available late in the growing season, when junior water users often ran out and when an additional application of water to high value crops such as sugar beets and potatoes might make the difference between a modest profit and debt. In stark financial terms, Parshall stated that irrigation generated almost one-third of the region's property value. This resulted in local, state, and federal taxes that could be invested in schools, infrastructure, and economic development.[53]

Knowing that C-BT detractors might argue that the nation was suffering from overproduction, Parshall turned that caution on its head by claiming that more water would shift agricultural production away from crops grown in surplus and toward crops not grown in sufficient quantities. For example, he argued that wheat, whose national supply had far outstripped its demand, was a crop of choice on the Piedmont only when water was in short supply. By contrast, domestic sugar beets, which demanded more water than wheat, supplied about one-quarter of the nation's sugar demand. According to Parshall's logic, increasing Piedmont water supplies would push farmers to grow more beets and less wheat, thus aligning the nation's agriculture more closely with consumer demand. This was a tendentious argument in 1937, given the quota system instituted by the Sugar Act. However, there was some accuracy to Parshall's conviction that more water would supply more beets since studies dating back to the 1920s correlated water increases with higher beet yields.[54] So, even if the Sugar Act regulated overall production, C-BT water could make farmers more efficient. Parshall concluded that more water on the Piedmont resulted in self-reliant, productive Americans who created real economic value in their region and beyond. In other words, the C-BT was an overwhelmingly good investment.[55]

To garner support for the project and sympathy for Piedmont farmers, Parshall employed some misleading water logic. To demonstrate the value of C-BT water, he compiled a dizzying array of statistics that compared the availability of water in a given year to crop yields, broken down by regions and irrigation companies. Despite available data sets extending back to the early 1900s, Parshall always chose the ten-year period from 1925 to 1934. This enabled him to contrast the year of highest water availability and greatest yield, 1926, to 1934, the least productive, lowest water year. When referring to 1926, Parshall consistently called it a year of optimal or even normal production when, in fact, 1926 represented an abnormally high water year. Then, when calculating money lost due to lack of water, Parshall displayed 1926 as a "break even" year, while all other years represented losses. From there, Parshall went on to display how the extra water brought by the C-BT would make the difference between disaster and solvency for farmers. He argued that previously, when farmers lacked sufficient water in a given year, they could buy it for prices that widely varied year-over-year but averaged $4.30 per acre-foot. However, Parshall and C-BT supporters claimed that the new source would run only $2.00 per acre-foot, concluding that costs would be paid back many times over in farmer productivity.[56]

While Parshall was an unqualified supporter of the C-BT in 1937, he left out any mention of water inefficiencies in favor of a narrative that left no doubt as to the region's water needs. Until 1937 Parshall had built his engineering career around shoring up the cracks in the South Platte's irrigation infrastructure. Fixing dams, controlling seepage, measuring water, surveying snow, and removing ditch-clogging debris were all attempts to make the greatest use of water that was already within the watershed. His research and advocacy, prior to the advent of the C-BT, emphasized that a watershed was a sealed system, and that the goal of the engineer was to create as much utility as possible within that system. In fact, before locals floated the C-BT idea, he generally tried to redirect conversations about trans-mountain

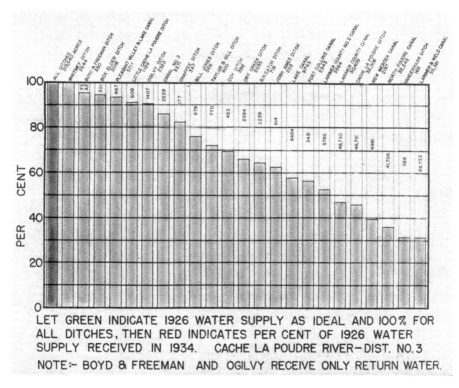

Fig. 24. Ralph Parshall compares water availability in 1926 and 1934. One of many diagrams employed by Parshall to promote the Colorado-Big Thompson Project. Parshall argued that full irrigation ditches, indicated by the line at the far left, should be the norm and that the extra water from the C-BT would make full ditches possible. Irrigation Research Papers, CSU Morgan Library Archives and Special Collections. Used by permission.

diversion.[57] In his mind, reversing water's flow during the Depression would result in costly water that farmers could not afford. The best answer was found in efficient use of the watershed. However, by the time the C-BT entered broad Piedmont consciousness, Parshall had witnessed the slow progress of his flume alongside the failure of mutual irrigation companies to shore up their creaky systems. The C-BT promised more water without requiring local districts to pay much for the works that would deliver it, since the region was only required to repay $25 million of the overall cost of a project that was budgeted to cost $50 million.[58] What Parshall knew was that the C-BT was a good financial deal for Piedmont farmers. So, for a time, he stopped beating the drum of efficiency and instead preached the gospel of water abundance.

If Parshall had reservations about the C-BT, Great Western Sugar did not. During the mid-1930s, its water rights remained insecure. In fact, during the 1934 refining campaign, the company's ditches ran dry, and it begged local irrigation companies for water to finish its refining operations.[59] More importantly, the company was conscious of the relationship between water availability, water security, and the number of sugar beets grown per acre each year. Increased water, company leaders reasoned, kept growers happy and its factories operating at capacity. Great Western employed this argument to great effect by publishing the work of researchers such as H. E. Brewbaker, who concluded that a little more irrigation late in the season would result in larger beets and more sugar.[60]

Further promotion by Great Western from 1936 to 1938 espoused much of the economic reasoning put forth by Parshall, peppered with homespun wisdom and couched in folksy stories. In its first promotional article in 1936, Great Western's president of Colorado operations exclaimed that the C-BT was the "biggest opportunity northern Colorado has had since the beginning of irrigation development in this country." He argued that the lack of trans-mountain diversion water had resulted in an average loss of $400 annually for each sugar beet grower. Put another way in a separate article, growers should expect three additional tons per acre with the new water available in their ditches. Moreover, when compared to other on-demand water sources in the region, the new water would be cheaper by at least 50 percent. Recognizing that many growers had begun pumping underground water during the recent drought, Great Western cited a CAC study claiming that C-BT water would recharge underground aquifers and thus, the wells that some growers relied on. Great Western even enlisted nationwide sources to make its case. In cooperation with other western beet sugar companies, its leaders convinced the National Broadcasting Company (NBC) to air a short radio program. Titled "Sugar Beets Tell the World," and transcribed in *Through the Leaves*, the program emphasized how sugar beets grown in the irrigated regions of the West stimulated the American economy, providing figures on grower income, railroad shipments, resources used in refining beets into sugar, and the varied ways it was consumed. As Great Western spun it, beet sugar flavored the American economy, and the C-BT could make it sweeter.[61]

Approved by Congress in 1937, signed off by Secretary of the Interior Harold Ickes, and finally completed by Reclamation twenty years later, the C-BT was a costly engineering marvel that eventually added 320,000 acre-feet of annual water to Piedmont supplies. Put in perspective, that bolstered water totals by 20 percent, the equivalent of doubling the entire average annual flow of the Cache la Poudre River. To bring the water to the eastern slope of the Rockies, engineers siphoned it from the Colorado River through four dikes and stored it in the massive 540,000-acre-foot Granby Reservoir. They then pumped it nearly two hundred feet uphill into the much smaller Shadow Mountain Reservoir, also constructed for the C-BT. Behind Shadow Mountain Reservoir, engineers dug a channel to move water into Grand Lake. The water then entered the Alva Adams Tunnel just one-quarter mile outside Rocky Mountain National Park, emerging three hundred feet east of park boundaries 13.1 miles later. From there, engineers diverted water into the Big Thompson River, as well as into a series of reservoirs, tunnels, and canals. Eventually the water plunged 2,900 feet and through nineteen additional dams until it flowed into the irrigation canals of Piedmont farmers. Though primarily intended as an irrigation project, the C-BT also supplied power and municipal water to regional communities, including Boulder, Longmont, Loveland, Fort Collins, and Greeley.[62]

Understanding why the C-BT was an economic boon for Piedmont farmers goes a long way toward explaining why agri-state engineers such as Parshall, as well as farmers and local water managers should occupy an important space in our understanding of water in the West. Parshall's initial skepticism about large water projects and transmountain diversion revolved around the ratio of benefit to expense,

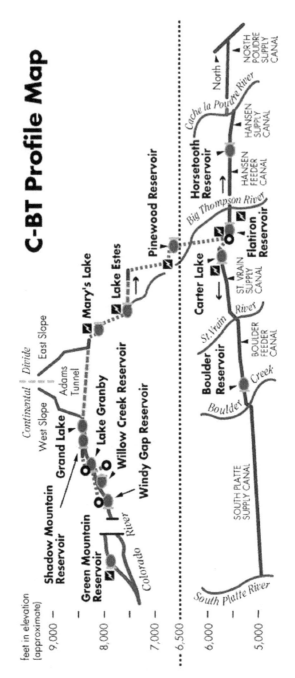

Fig. 25. Colorado-Big Thompson Project elevation profile. Northern Colorado Water Conservation District. Used by permission.

and $2.00 per acre foot was a great deal when compared to what users might pay for water on demand otherwise. In addition, the $25 million price paid over forty years was palatable as well, especially when considering that the total costs of the C-BT, when finally built out in 1957, were approximately $150 million. The same farmers, irrigation companies, local boosters, and sugar company who were primarily responsible for crafting the C-BT and gaining Reclamation support, obtained an overwhelming bargain. But their success was not just in the price tag. Despite employing federal dollars and expertise, Piedmont water users gave up none of their autonomy to the federal government. The C-BT's authorizing legislation contained no provisions for opening new lands to farming or limiting the amount of water that individual farmers could divert. Even though the average Piedmont farm was less than one hundred acres when the project was approved, C-BT legislation opened the door for successful farmers and agribusiness to consolidate smaller farms into larger ones without concerning themselves with access to Reclamation water on demand. In the end, the C-BT provided more water for a well-developed agricultural region while doing nothing to encourage more farmers to settle there. If anything, since water was distributed through mutual irrigation companies, the C-BT provided a road map for enriching those who already possessed water rights.[63]

Distribution of C-BT water also points to a delimited federal power. The Bureau of Reclamation managed all water coming from the initial point of diversion until it found its way into irrigation company canals. From that point forward, the bureau relinquished control over the distribution of C-BT water to the Northern Colorado Water Conservation District (NCWCD). Formed in 1937, the NCWCD functioned to disperse C-BT water to irrigation companies on demand. Once irrigation companies ordered a specified amount of water, Reclamation's only role was to deliver it. Decisions on how to disburse it were entirely in the hands of irrigation companies and the conservation district they formed.[64] Ralph Parshall was fully aware of these facts when the Bureau of Agricultural Economics hired him to write the economic rationale for the C-BT. Though an employee of the federal government, he viewed the project through much the same lens as he viewed his namesake flume. Whether through efficient measurement or increased quantity, both the Parshall Flume and the C-BT aimed to make water more available and secure to a broader swath of farmers. Parshall was effective to the degree that he made common cause with them.

The C-BT is a reminder of the critical yet bounded power of engineers such as Ralph Parshall. While Parshall's irrigation work for the Bureau of Agricultural Engineering and his deep roots on the Piedmont made him the ideal choice to craft the C-BT's economic rationale, his influence rarely penetrated the politics or culture of water management. To a large extent, the same leaky and inefficient infrastructure that diverted the region's water remained in place. What held true for C-BT water also held true for the Parshall Flume. Despite the clear advantages in water distribution made possible by the Flume, its adoption on the Piedmont proceeded in fits and starts. As an agri-state engineer, Parshall's greatest influence came when he employed technocratic expertise to provide scaffolding for perceived local needs, and on a timetable that he rarely influenced.

The C-BT also helps us to understand the big changes afoot in Piedmont agroecology. At its most fundamental level, the C-BT represented a successful attempt to unmoor Piedmont agriculture from the region's limited water supply, enlisting agri-state engineers as agents who would construct and maintain the infrastructure. Ralph Parshall provides an example of a bridge to the transition. His life's work illustrated an unrelenting desire to understand and engineer how a defined watershed could benefit the greatest number of people living within it. When Parshall shifted his energies to researching and writing a full-throated endorsement of the C-BT, he implied that, given the state of irrigation politics on the Piedmont, the region's water resources would never completely supply the demands of regional water users. Parshall, like his agri-state colleagues who supported and built the C-BT, employed technocratic scaffolding and physical energy to reengineer one landscape with the resources of another.

Mechanized sugar beet cultivation illustrates a different kind of extra-local transformation. The decline in contract labor that began in the 1930s did not so much represent a move toward reducing reliance on imported workers as it did a shift toward another kind of energy dependency. The tractors and their host of cultivation and harvesting attachments that replaced humans and horses were built with materials extracted from far-flung locales and assembled outside the region. For agri-state and industry scientists working on the Piedmont, mechanization was part of a larger reorientation of research agendas and an expansion of their clientele. Prior to the 1930s, the lion's share of agri-state efforts was targeted at concerns such as soil health, water usage, and aligning animal husbandry with crop rotations, all of which supported the aims of farmers and industry. Pivoting toward mechanization—energized in part by the Sugar Act—was part of a larger shift toward growing yields and reducing costs per acre. Once machines slashed cultivation costs, agri-state scientists in the 1940s could turn to organic chemicals, the next agricultural frontier, which promised to eradicate organisms that eroded production goals. These would threaten another pillar of Piedmont agroecology.

Once turned loose, controlling the direction of the changes energized by the Depression proved more difficult than bringing them about. While mechanization reduced costs in the beet sugar industry, it also opened the door to other kinds of crops that were more easily adapted to tractors, cultivators, and harvesters. Ironically, one crop in particular, corn, benefited from the increased water security offered by the C-BT, while threatening to replace the sugar beet pulp that was a foundational Piedmont livestock feed. By slashing the labor of humans and animals and shifting toward trans-mountain water diversion and energies derived from machines and fossil fuels, Piedmont farmers, agri-state scientists, and industry leaders destabilized Piedmont agroecology, yielding an unclear future.

6 Building the Petrochemical Paradigm

> One either recognizes all the intertwining webs around the cowpie or one drifts gradually into assembly line economics.
>
> —GENE LOGSDON

In 1930 two Weld County farmers made fateful decisions that radically altered Piedmont agroecology. William (W.D.) Farr, a twenty-year-old Greeley native and recent college dropout, approached his father, Harry, with a request to feed beef cattle full-time on his family's land. This was unprecedented on the Piedmont, and Harry was skeptical. In the Farr family, and throughout the region, fattening cattle was a seasonal side business whose greatest benefit was in its contribution of manure to soil health and crop diversity. In addition, the market for beef cattle collapsed during the 1920s. And besides, when it came to raising livestock, the Farr family achieved greater success marketing its lambs. In 1930, W.D. saw an opportunity when Great Western Sugar offered his father free beet pulp if he would agree to lease a large plot of land east of Greeley. The additional land and extra feed gave Harry a low-risk way to placate his son. The junior Farr carved out a small plot of the leased land to build a pen for feeding cattle year-round. In 1930 he packed it with 132 animals.[1]

That same year, Warren Monfort had the same idea. A veteran of World War I and a former teacher from Illinois, Monfort moved to Greeley in 1921, where he leased eighty acres of family farmland. Like the Farrs and most of his Piedmont neighbors, Monfort raised sugar beets for Great Western Sugar, as well as a rotation of supporting crops, while fattening modest quantities of cattle in the winter. While living in Illinois, Monfort observed how corn farmers employed their crops in the winter to finish cattle, selling them to one of the Big Five Meatpackers in the spring.[2] Noting corn's decided feeding advantage over sugar beet pulp, Monfort wanted to do the same. However, since corn was a marginal Piedmont crop, he imported it from the Midwest. In 1930 Monfort started with a modest eighteen animals. Five years later, he penned 1,500 cattle.[3]

As W. D. Farr tells it, the business model for the full-time cattle feeding industry crystallized one day in the early 1930s when he and Warren Monfort sat together on a fence rail waiting for a train that would freight their lambs away to slaughter. Comparing notes on their cattle-feeding efforts to date, they came to several conclusions. The two agreed that commercial cattle feeding must be a year-round operation. Prevailing Piedmont practice timed cattle purchases and sales to coordinate with planting and harvest. This resulted in purchasing feeder cattle when prices were high and selling finished cattle in a glutted market defined by low prices. Midwestern corn farmers operated on a similar timeline. Monfort and Farr observed that rationalizing the cattle market required buying and selling throughout the year. The two were particularly interested in Colorado's tourist market, centered on towns such as Estes Park—gateway town to Rocky Mountain National Park—where restaurants and hotels bought directly from cattle feeders.[4] Monfort and Farr realized, however, that their cattle-feeding model could not thrive within the confines of Piedmont agriculture since the principal feed, beet pulp, was only available during the winter and was not scalable to meet their ambitions. So, they purchased train-car loads of Midwestern corn. In this, they went against the grain of longstanding Piedmont logic that argued that the benefits of feeding corn could never justify its shipping expense. While this had been true in the past, recent advances in mechanizing corn cultivation alongside declining demand during the Depression glutted the market, enabling aspiring Piedmont cattle feeders to buck prevailing wisdom.[5]

Fig. 26. Photograph of W. D. Farr. City of Greeley Museums. Used by permission.

Fig. 27. Photograph of Warren Monfort. City of Greeley Museums. Used by permission.

While there is no evidence that Farr and Monfort discussed Piedmont agroecology that day on the fence rail, it is clear that their model for commercial cattle feeding was a blueprint for severing existing economic and biological relationships. The region's keystone beet sugar industry was built on a series of interdependencies. Sustainable harvests of sugar beets relied on extensive crop rotation and fattening livestock seasonally for their nitrogen-rich manure. Healthy soils supported beet cultivation, which sustained profits for Great Western Sugar. In turn, the company supported its growers with modest financial returns and beet pulp, a cheap and effective livestock feed. The capitalist ethos of simplifying landscapes for rational economic gain was restrained by the need to support a complex agroecology with locally available resources. By contrast, Monfort and Farr's corn-dependent, year-round commercial cattle feeding model operated on an alternative logic. It assumed that the greatest value of a piece of land was in its ability to grow the maximum quantity of high-priced beef at the lowest cost in the shortest period. Under that logic, soils did not consist of a diverse and interconnected set of organic relationships, but a constellation of matter capable of supporting animal heft. Further, while animal feeds could certainly be supplied locally, this was not essential. The free market, unmoored from local agroecology, governed decision-making. Consequently, if Warren Monfort and W. D. Farr's corn importation gamble paid off, it would prove that sugar beet byproducts were unnecessary for industry success. And, what about all of that nitrogen-rich manure? The race to fatten cattle fast certainly produced copious quantities capable of fertilizing fields. However, for those who chose to go whole hog into commercial feeding, cattle manure was waste, not life. Farr and Monfort's model made no allowance for balancing livestock quantities with support for living soils. In short, commercial cattle operated on an alternative agroecological and economic logic than the prevailing one on the Piedmont during the 1930s. The two could interact but not interdepend.[6]

The unraveling of Piedmont economic and agroecological relationships from the 1930s through the 1950s also hung on critical technological developments in chemicals and fertilizers. Eventually embraced by farmers and feeders alike and promoted by agri-state scientists, commercial chemicals and fertilizers were of limited utility when Farr and Monfort built their first feedlots. Prior to 1940, Piedmont farmers' episodic deployment of these synthetics pointed to an agroecology reliant primarily on nature's energies. Farmers applied insecticides, such as the arsenic-based Paris Green, to crops in reaction to insect infestations rather than as a set of tools for reordering farming practice.[7] Evidence also shows that Piedmont farmers used commercial fertilizers sparingly. Of the three most critical plant nutrients, nitrogen and phosphorous were cycled into the soils through longstanding farm practices, and commercial potassium was almost never used since Piedmont soils had long been storehouses of that nutrient.[8] The seed treatments and disinfectants employed by the mid-1930s provide the only noteworthy examples of how Piedmont farmers systematically integrated chemicals into cultivation. This change largely resulted from research performed by petrochemical companies, such as Dow, Dupont, and Shell, who networked with agri-state scientists to perform field trials at land grant colleges, such as Colorado Agricultural College (CAC).[9]

That network metastasized after World War II in response to a chemical revolution that had been bubbling below the surface during the 1920s and 1930s. Whereas, prior to World War I, American chemical companies formulated pesticides and fertilizers with inorganic heavy metals, such as arsenic and lead, the postwar compounds they synthesized employed carbon molecules extracted from fossil fuels. These organic insecticides, herbicides, and fungicides were capable of eradicating specific pests, weeds, and soil fungi without harming desired crops and livestock. Chemical companies who possessed large stocks of volatile elements, such as chlorine; nitrogen; magnesium; and mercury, after World War I, formed varied relationships with the fossil fuel industry that undergirded the development of new compounds. Agriculture presented one of several outlets for their organic chemicals and land grant colleges were eager to perform field trials before companies rolled out the new biocides.[10] Agro-industry also fused petrochemical energies using the revolutionary Haber-Bosch Process, which synthesized ammonia (NH_3), a compound capable of providing easily assimilable nitrogen to plants. While few commercial outlets for synthetic nitrogen existed in the United States prior to World War II, the technology had the potential to supplant existing agroecological relationships by replacing organic processes of soil restoration with synthetic ones.[11] Witnessing these changes, ecologist Barry Commoner observed that the prewar explosion of synthetic fertilizers and petrochemicals produced "sciences capable of manipulating nature—of creating for the first time on earth, wholly new forms of matter." But it would not be until after World War II that their influences were felt. Those would have a transformational effect on Piedmont agroecology.[12]

Chemicals and Cattle

Of the host of organic compounds pouring out of chemists' labs and onto the Piedmont landscape after World War II, dichlorodiphenyl-trichloroethane (DDT) received the most advanced publicity and the greatest use. DDT is one of several organic compounds classified as chlorinated hydrocarbons, due to its synthesis of chlorine, hydrogen, and carbon molecules. Chemists synthesized DDT in the 1870s, but it was largely left on their shelves until the Switzerland-based Geigy Chemical Corporation used it successfully to eradicate Colorado potato beetles during the 1930s. Geigy representatives approached the USDA in 1942 with the compound, suggesting it could possess wartime value. Already searching for chemicals that might replace existing crude methods for protecting troops and civilians from diseases, scientists quickly recognized the utility of DDT's insecticidal properties. Initially used to quell a typhus outbreak in Italy in 1943 and 1944, the U.S. Army employed the biocide to greatest effect protecting troops against malaria in the Pacific Theater in 1944 and 1945. Word of DDT's insecticidal prowess spread quickly through the farming community, which hoped that its wartime success could translate to agricultural bounty. Despite concerns by the FDA, the USDA and the U.S. Public Health and Fish and Wildlife Services about DDT's toxicity to humans, other animals, and soils, the U.S. War Production Board allowed chemical manufacturers to move forward with production and sales in July 1945. Field tests conducted by chemical companies, USDA experiment stations, and agricultural colleges revealed that DDT's insecticidal properties generated significant yield increases in tomatoes, onions, and potatoes. DDT quickly replaced rotenone as the go-to insecticide for protecting feedlot cattle from flies, lice, bedbugs, and fleas, due to its potency and persistence, resulting in more insects killed with fewer applications. Within just two years of the war's conclusion, DDT was sprayed on crops, stock, domestic pets, wetlands, urban neighborhoods, and even incorporated into wallpaper formulations. In 1948 there were twenty million pounds of DDT available in the United States. Ten years later that number increased more than sevenfold.[13]

Piedmont cattle feeders were especially well situated to embrace DDT. By 1952 over two-thirds of all cattle in Colorado were confined on feedlots and Weld County led the nation in number of cattle on feed.[14] In the feedlot environment, energy spent fending off insects was energy diverted from gaining weight. The physical layout of the feedlot exacerbated the problem. Put simply, feedlots were pest magnets. When the same animal species—or crop species for that matter—was raised in one locale year after year, their natural enemies concentrated in that same location. Some insects that target cattle, such as flies, evolved to hatch their eggs in cattle feces. Though prewar cattle were generally dipped, sprayed, or dusted with formulations of rotenone, that inorganic pesticide repelled insects for only a few days. By contrast, DDT's stability enabled it to control pests for three weeks, minimizing the time and labor of repeated applications.[15]

Knowledge of DDT's capacity to control insects that targeted cattle often originated from the same cast who demonstrated its utility on crops. In 1946 CAC extension entomologist Gordon Mickle, along with colleagues from Colorado and Wyoming, joined officials from the USDA's Bureau of Animal Industry and Bureau of Entomology and Plant Quarantine (BEPQ) to plan demonstrations for the follow-ing year. To attack a persistent Piedmont lice problem, they dipped fifteen thousand cattle in DDT solutions in 1947, advertising the demonstration on a Greeley-based radio station. They modeled these DDT demonstrations after those being performed in Kansas in 1946, where BEPQ worked with Kansas State's experiment station to treat thousands of cattle. The cooperating agencies boasted that two men standing on catwalks could spray over three hundred cattle in a short space of time by herding them though narrow chutes while spraying their sides and backs. The operators also sprayed barns, sheds, and other structures frequented by cattle. The goal was complete control of flies, lice, fleas, bedbugs, and mosquitoes. According to the experiment's authors, cattle treated with DDT gained seventy-three more pounds while in confinement than those that were untreated, at the cost of only four to ten cents per head. Inspired by that success, the Kansas City Stockyards began a DDT spraying service in 1946. For six dollars an entire train-car load could be sprayed before their journey to a feedlot. Many of those feedlots were on the Piedmont. Consequently, the region's cattle feeders received animals that were remarkably insect free—courtesy of DDT—and it did not take long for Piedmont cattle feeders to become DDT converts. So, whether it originated from industry, experiment stations, chemical retailers, agencies of the USDA, or the bodies of cattle themselves en route to a multitude of feedlots, the DDT gospel spread rapidly. As the confined cattle feeding industry grew on the Piedmont in the 1940s and 1950s, DDT's presence grew with it.[16]

Agri-state veterinarians echoed the calls of their entomologist colleagues for integrating chemicals into livestock management. In 1947 new CAC extension veterinarian Albert Goodman carried the message from the USDA and experiment stations to feedlot operators. Arguing that insects and parasites alike were enemies of sanitation, he told them to spray DDT wherever they could be found. This not only included on cattle and on buildings but also on other farm animals, on damp piles of straw, and wherever moisture might attract mosquitos. Going further, Goodman instructed them to spray DDT on manure piles, reclassifying this essential soil replenisher as the enemy of farm sanitation. Like his colleagues, Goodman was clear to link hygienic benefits with economic ones, stating, "Controlling lice, flies, grubs, and other parasites is one of the livestock producers' quickest and cheapest means of increasing our nation's needed supplies of meat, milk, and leather." According to Goodman, insects and parasites damaged livestock hides equivalent to two million pairs of shoes and twelve million pounds of meat.[17] Goodman's message to livestock growers, just two years after FDA approval of DDT, offers evidence of how rapidly Piedmont feeders adopted the pesticide and the broad spectrum of purposes to which it was applied. Concentrated animal feeding, chlorinated hydrocarbons, and the institutions that supported their use mutually reinforced each other.

DDT's rapid proliferation presented well-documented dangers to organisms and ecosystems. Its chemical stability meant that wherever it accumulated, it remained for decades. When DDT sprays and dusts were washed from cattle and barns, or when DDT-laced manure was spread over fields, much of it bioaccumulated in soils. Based on DDT's extensive employment in agriculture from 1945 into the early 1960s, up to forty pounds of DDT per acre accumulated in the croplands of the United States. However, since it does not penetrate deeply into the soil, precipitation drained most of the remainder into the watershed. There, fish were especially susceptible since chlorinated hydrocarbons interfered with respiration at the gill membranes, resulting in death by suffocation. DDT's persistence, paired with its prevalence in agriculture, translated to accumulation in humans. Federal scientists testing DDT in advance of World War II learned that the body stored chemical residues in fatty tissues. DDT massing in the fats of mammals soon found its ways into their milk, making the dairy industry a source of transmission into the human body. By 1963 the mean quantity of DDT present in human fatty tissues had risen to twelve parts per million, and the typical meal contained two parts per million.[18]

The omnipresence of DDT provides the starting point for understanding resistance, one of the most important consequences of its use. By 1948 the petrochemical was no longer able to kill certain insects at the same dosages. Consequently, consumers chose to use more DDT until either cost or hazard made such use prohibitive. Resistance to DDT was simply one example of species adaptability to environmental change. No individual insect within a species possesses the exact same genetic makeup as another, and so, no single individual represents the norm. According to ecologist Robert Rudd, biocides are simply agents that "eliminate the more susceptible individuals." Those that survive, and are therefore resistant to the chemicals, pass on that trait to future generations. Over time, as susceptible individuals are eliminated, resistance to particular chemicals becomes a dominant trait within a species. Entomologists had observed insect resistance to chemicals in the past but were astonished by how quickly certain species developed resistance to DDT. In 1948, when researchers confirmed that certain species of flies had developed resistance to DDT, twenty million pounds of the product were manufactured. As it accumulated in soils, waters, and bodies, production increased in the next decade to 145 million pounds. In 1963 at least thirty insect species were resistant. As these statistics suggest, initial discoveries of resistance did not present a cautionary tale, but rather an impetus to use more.[19]

On the Piedmont, where DDT had become an indispensable tool for cattle feeders, insect resistance and DDT limitations encouraged users to embrace an array of chemicals. One insect that failed to succumb to DDT was the cattle grub. The term refers not to a species but to the larval stage of a heel fly. In that stage, larvae crawl down the hair and penetrate cattle skin, causing considerable irritation. Agricultural scientists found benzene hexachloride (BHC), another chlorinated hydrocarbon, effective at killing cattle grubs. So, cattle feeders applied both DDT and BHC. Soon, BHC-resistance developed among several insects. Consequently, during the early 1950s, cattle feeders gradually increased their dosages of both chemicals. They also added other chlorinated hydrocarbons, such as toxaphene, lindane, and chlordane to their arsenal. Resistance to those compounds led agri-state researchers and the feeding industry to seek relief through organophosphates, another class of organic compounds. Less than ten years after the introduction of DDT into the cattle industry, feeders were convinced that multiple classes of chemicals were essential to their operations.[20]

Agri-state scientists not only paved pathways for the introduction of the new chemicals but promoted their indispensability. USDA entomologist Arthur Lindquist offers a telling example. Starting his career as an entomologist for Great Western Sugar during the 1920s, Lindquist was hired as a junior entomologist for the USDA in 1931. During the 1940s he conducted research on controlling malaria with DDT, oversaw the DDT spraying demonstrations in the Kansas City Stockyards, and performed lab experiments demonstrating how flies developed resistance to the compound. Lindquist's research showed how insecticides killed parasites that preyed on livestock, eventually developing systematic treatments used to control cattle grubs. Promoted to chief of the Insects Affecting Man and Animals Research Branch of BEPQ in 1953, Lindquist was as informed and influential as any agri-state scientist working at the intersection of biocides and livestock. Lindquist argued that feeders would have to depend on a host of synthetics into the future—from multiple classes of chemical compounds—to kill all manner of animal parasites. He further claimed that, as a consequence of resistance, experiment stations and the USDA should allocate more resources to finding new and useful compounds and replacing those with declining effectiveness. To justify the employment of taxpayer dollars on applied biocide research and field trials, Lindquist cited dramatic increases in American meat consumption, arguing that affordable meat was inextricably linked to a growing range of petrochemicals and that consumer demand determined agricultural practice.[21]

Lindquist's argument demonstrates how completely chemical control of unwanted pests and economic imperatives had come to dominate agri-state science. Each new chemical added to the cocktail necessary to kill cattle-targeting insect pests presented an opportunity to question whether synthetics were the solution. There is little evidence however that researchers considered other options. For example, cultural and biological control might have included reducing the number of cattle in pens, researching organisms that preyed on cattle pests, or searching for methods to effectively separate cattle from the manure that provided insect breeding grounds.[22] That chemical alternatives were generally not on the agenda suggests that the relatively low cost of synthetics and the economic value of confined cattle feeding trumped alternatives.[23] Moreover, confined cattle feeding was analogous to monocropping since feedlot operators grew the same species on one plot of land year in and year out. Feeders chose animals for their rate of economic return. Using the land for another purpose, no matter how ecologically healthy, was akin to reducing one's income. As long as biocides enabled feeders to monocrop cattle, they continued to do so. Consequently, by arguing that an ever more exotic spectrum of chemicals was necessary into the foreseeable future, Lindquist implied that alternatives to synthetics were no longer feasible.

While the demand for beef and proliferating biocides increased the appeal of commercial cattle feeding, by themselves organic chemicals did not overthrow prevailing practices on the Piedmont. In the first decade after World War II most farmers incorporated synthetic chemicals into livestock management, while still cultivating the same diverse array of crops that had predominated for decades, feeding cattle to support soil health and the farm economy.[24] Iliff farmer and former Colorado state representative Don Ament describes it this way, "These were diversified operations. . . . So you sold beets, raised corn, some beans, some alfalfa, and you had to have the cow/calf thing going and you had to have that for the income. . . . [We] always fed only what we grew on the farm."[25] Chuck Sylvester, who grew up near Greeley and became the director of the annual National Stock Show in Denver, agreed. He observed that "every farm had a small feedlot and fed out cattle and steers." Peers of Ament and Sylvester recall just how essential soil health remained in the decade following World War II. Describing the landscape around the small town of Mead growing up, former Rocky Mountain Farmers' Union president Kent Peppler recalls that every quarter-section (160 acres) had at least one field of alfalfa that was being used to feed livestock and to restore nitrogen to the soil. His neighbor Richard Seaworth concurs, adding that in order to feed cattle and grow sugar beets, his field required three years of alfalfa out of seven. As for sugar beets and their by-products, they remained essential. As a Piedmont high school student in the 1950s, Dick Maxfield recalls growing sugar beets and making trips to Great Western's factory in Fort Collins to haul beet pulp, describing how his family viewed it as one of the best cattle feeds available.[26] While there is no question that chemicals were becoming commonplace in the decade after World War II, agricultural practices suggest that crop farmers on the Piedmont expected that the new synthetics would augment existing practices, not remake them.

Beets and Chemicals

Ultimately, the new chemical paradigm could not harmonize with previous practices because synthetics promised more than cheap efficiency; they required a complete philosophical reorientation. As the historic keystone Piedmont crop, sugar beets offer the logical place to peel back the chemical layers of this reality. With the advent of segmented seed during World War II and the long-hoped-for mono-germ breakthrough in 1951, sugar beet growers mechanized their operations, replacing hand laborers with a bevy of cultivators and harvesters attached to tractors, which had formerly been of limited utility. As with commercial cattle feeding, the chemical industry, beet growers, and agri-state scientists anticipated the increased efficiencies that organic biocides offered. Initially, they hoped that DDT would be an insecticide panacea. Their hopes went largely unfulfilled. Field tests conducted by Great Western plant pathologist Albert Isaksson, aided by research at CAC, found that DDT resulted in only limited success, controlling a select group of insects that fed on beet seeds and mature plants. Isaksson's colleagues at the USDA's Division of Sugar Plant Investigations (SPI) and BEPQ conducted multiyear studies that confirmed DDT's insecticidal weaknesses, while also finding evidence that it sometimes killed plant foliage.[27] Again, DDT's failings led industry and agri-state scientists to search for and test a bevy of new biocides.

Studies conducted by agri-state scientists and the beet sugar industry show that Piedmont beet growers were more concerned with eradicating organisms that resided in the soil than in the air. While the field of entomology had emphasized killing insects since the early twentieth century, there was no analogue in weed science.[28] Prior to World War II, eliminating weeds was performed by hand or machine. The selective nature of the new organic compounds literally made weed science viable, since the right compound killed certain weeds and left food crops unharmed. Herbicidal frontiers pushed industry and agri-state institutions to locate and train experts, leading to the introduction of new products.[29] Since mechanized cultivation of sugar beets was in its infancy and migrant workers still completed most weeding by hand, growers were keen to find cheap and effective alternatives. Dichlorophenoxyacetic acid (2,4-D), an organic compound that received great herbicidal attention during World War II—not unlike the promise of DDT as an insecticide—presented the most possibilities since it killed a broad spectrum of weeds. However, field trials with sugar beets showed that 2,4-D also killed beet seedlings and seeped into ditches. More problematic, the new compound could potentially infect the fields of unsuspecting neighbors simply through the ubiquitous act of irrigation.[30]

The failure of 2,4-D opened herbicidal floodgates for Piedmont farmers. The same rotations that made sugar beets a sustainable crop also presented cultivation quagmires in May and June when seeds germinated and plants were at their most vulnerable. Their chief competitors were the previous years' crops. The same oats, alfalfa, and pasture grasses that were cultivated one year, dropped seeds that became weeds in succeeding years, interfering with sugar beets and other crops. Consequently, industry and agri-state scientists evaluated the utility of herbicides for killing food crops with the audacity to germinate during inconvenient periods. Beginning in 1948, G. W. Deming, a USDA agronomist working out of CAC in Fort Collins, tested another class of industry petrochemicals called carbamates, yielding several compounds that killed grasses, oats, barley, and wheat. Agri-state scientists and peers at chemical companies and Great Western Sugar followed up on these with their own field trials that synthesized carbon with elements and compounds such as chlorine, phosphorous, and various acids. In the ten years following World War II, a diverse network of industry and agri-state scientists established a robust program for herbicidal research. Their frenetic pace was animated first by the failure of the most common agro-chemicals and second, by a narrow belief that the solutions to their weedy problems lay in concocting more chemicals.[31]

Weeds and untimely food crops were not the only earthbound organisms that competed with crops for water, sunlight, and nutrients. Subterranean organisms such as nematodes and wireworms preyed on plant roots, robbing them of their vigor or killing them outright. Consequently, Piedmont farmers increasingly adopted a growing selection of fumigants and fungicides. Applied prior to planting, these biocides indiscriminately killed a host of organisms that constitute soils. While farmers applied fumigants such as chloropicrin to Piedmont crops prior to 1940, petrochemical companies commercialized a wider and more toxic array after the war. Among the chlorinated hydrocarbons offered, Piedmont farmers adopted Shell Corporation's D-D product and Dow's Dowfume most readily. To achieve efficient soil penetration, farmers deployed tractor attachments that injected soils with the chosen fumigant, usually at the rate of ten to sixty pounds per acre, depending on the degree of death desired. Subterranean injection was critical since these biocides were acutely toxic when airborne. Once applied, farmers waited an average of two weeks for the fumigant to dissipate prior to planting. For most Piedmont farmers, fumigants were inexpensive tools for killing invisible pests.[32]

Selective herbicides and especially fumigants came at an agroecological cost, requiring farmers to replace soil nutrients synthetically that had formerly been supplied organically. By killing organisms in the soil that preyed on crops, farmers disrupted a living network of life that supplied critical plant nutrients. To fill that vacuum, Piedmont farmers embraced commercial fertilizers as never before. Economically speaking, commercial fertilizers and organic biocides were complementary goods. Field trials and demonstrations conducted by researchers and extension agents at the USDA and CAC, as well as by Great Western Sugar, helped farmers gain knowledge of concentrated forms of synthesized nitrogen and phosphorous. By 1950 anhydrous ammonia, a gasified form of nitrogen, became available at various terminals throughout the Piedmont. Inserted into irrigation ditches and injected into soils, anhydrous ammonia became one of the most convenient and efficient ways of replacing nitrogen lost due to fumigation and changing farm practices.[33]

Statistics bear out the changes. While figures on commercial fertilizers used throughout the Piedmont are spotty prior to 1954, Great Western Sugar collected data on biocides and fertilizers used by its growers much earlier. In 1946 its growers applied twenty pounds of phosphorous per acre. By 1953 they spread over fifty pounds per acre. Since nitrogen had historically been supplied by manure and alfalfa, most beet growers were slow to adopt commercially produced synthetic varieties, applying only five pounds per acre in 1946. In 1953 that figure rocketed to forty pounds per acre. Growers were encouraged in the trend as beet yields rose noticeably during the same period.[34] The following year, the Colorado agricultural census began recording purchases of all commercial fertilizer used in the state, updating the data every five years. The first two censuses emphasized continuity in sugar beets and similar trends throughout the Piedmont. By 1959 beet growers applied nearly one hundred pounds per acre of phosphorous and over sixty pounds of nitrogen. During the same period, Piedmont farmers doubled their corn acreage, a crop that used 46 percent of the state's available nitrogen. In Colorado, purchases of commercial fertilizer increased by 57 percent from 1954 to 1959 with Piedmont farmers employing approximately half of all commercial fertilizers in the state.[35]

Chemicals and Agri-State Science

While the development of these biocides and fertilizers on the Piedmont owed much to the collaboration between agri-state scientists and industry, experiment station researchers and extension agents at CAC played the greater role in their integration. With the barrage of new petrochemicals available to them, farmers were hard-pressed to select from among multiple options to augment or replace prevailing practices. For example, if a farmer selected a fumigant such as Shell's D-D, the manufacturer offered nothing more than information on how and when to apply the biocide, alongside cautions and recommendations as to its usage. The same farmer required advice on which commercial fertilizers could replace nutrients lost to fumigation, and how these changes would impact the use of traditional soil-restoring manures and legumes. Moreover, each product that replaced or augmented prevailing practices required a host of complementary adjustments. This is not to suggest a poverty of knowledge on the part of Piedmont farmers, but rather to punctuate the speed by which a new knowledge economy emerged to replace the old one. In the initial years after the war, Piedmont county extension agents scrambled to transmit accurate information on biocide effectiveness, the application of various fertilizers, and how crop rotations might be adjusted to account for the changes.[36] By the mid-1950s, CAC's experiment station at Fort Collins published and distributed regular reports that enabled farmers to integrate the new chemicals into their cultivation practices. Reports included updated knowledge on available biocides, including their various formulations, how to mix and apply them, which crops to use them on, and some of the dangers associated with each one. They also gave farmers some insight into the future of synthetics and fertilizers by mentioning which ones were experimental and sharing the progress of biocide and fertilizer research. Their resulting publications and demonstrations functioned both as how-to guides and shopping lists for the synthetic farming frontier, offering a road map for agroecological transformation. By the mid-1950s, agri-state scientists such as those at CAC did more than just perform field trials for agro-industry; they systematized farming to fit the new chemical paradigm.[37]

Though farmers could have chosen to ignore the agricultural expertise extending from CAC, there is little question that, by the 1950s, its complex of scientists and demonstration agents were deeply embedded in Piedmont agriculture. Examples can be found in beet sugar industry literature where Great Western Sugar's monthly grower magazine had been publishing experiment station research since its inception in 1913. They can be found in the ditches that ran through the region's farms and feedlots where, by the mid-1950s, every engineered mile offered evidence of the work of CAC and USDA water engineer Ralph Parshall and his colleagues. Moreover, since the early twentieth century, Piedmont agriculture was peppered with CAC graduates. During a series of 2019 interviews with more than a dozen individuals possessing multigenerational roots in Piedmont farming and cattle feeding, CAC emerged frequently. Without exception, interviewees praised the agronomic research at the university. Cattle feeders never failed to mention the college's aid in diagnosing and treating bovine diseases, guiding them through complex regulations, or doing cooperative studies. Kent Peppler, a fourth-generation Piedmont farmer whose family began cultivating sugar beets in 1909, stated it plainly: "[CAC] was the Bible for us. We trusted their research and their advice and, as far as we were concerned, if [CAC] said it, that's the way it was. . . . We're betting everything that [CAC] is right." Though Peppler's effusive language may come across as hyperbole to some of his peers, it accurately portrays the value that Piedmont farmers ascribed to the university's science and advocacy.[38]

Despite CAC's depth of commitment to Piedmont agriculture, university researchers and their agri-state colleagues eschewed an essential function when they failed to question whether the changes wrought in the postwar period were in the best interests of Piedmont farmers. From a short-term economic standpoint, the new biocides and fertilizers worked as inexpensive forms of energy that killed undesirable organisms while supplying essential plant nutrients. Despite this, the work of land grant colleges such as CAC and its agri-state allies was short-sighted and reflected ignorance of their core mission. Land grant scientists were charged with employing institutional knowledge of local and regional agriculture to advocate for the long-term health of farmers and their lands. Yet, in their efforts to provide farmers with more tools for increased yields, they focused most of their research on agro-industrial field trials and then directed their extension agents to help farmers integrate the resulting products. While these trials for industry were common on the Piedmont since the 1920s, during the postwar period they over-whelmed other potential research agendas. Those included studies of plant toxicology, insect ecology, and adjusting crop rotations to accommodate for nutrient deficiencies. By failing to exercise sufficient skepticism—a central principle within any academic enterprise—and conducting original research to pursue that skepticism, land grant colleges such as CAC compromised their function as guardians of farmers and farmland. As a consequence, the intellectual center of agricultural research moved to distant laboratories where corporate profit informed research imperatives.[39]

To be sure, experiment station scientists and extension agents did seek to protect farmers from the harms of the new biocides, but it was a narrow form of advocacy. As historians J. L. Anderson and David Vail have argued, land grant scientists implored farmers to use the new chemicals as directed and to follow instructions provided by manufacturers. Anderson goes further to show how much of the responsibility for chemical overuse lay with farmers who often defied manufacturer instructions by applying excess biocides. Vail convincingly demonstrates that many of the pilots who sprayed chemicals from the air used the applied science coming from agri-state researchers as a form of restraint against applying biocides irresponsibly.[40] Though their research provides needed counterpoints to works that villainize land grant scientists and industry professionals, it is also clear that the agri-state complex uncritically embraced the new chemical paradigm.

The literature penned by CAC scientists aimed at farmers during the 1940s and 1950s developed a chemical-centered narrative of its own. Bulletins typically began with a nod to the value of good crop rotations and maintaining farm tools, then continued their larger discussion of biocides and/or fertilizers with strongly worded directives to follow manufacturer instructions to avoid residues on crops for human consumption and to exercise caution in methods of application. The bulk of these publications, however, advocated for the use of specific biocides and fertilizers on crops, livestock, pastures, and buildings. While these guides performed the function of helping farmers choose from among an array of options, their overwhelming emphasis on biocides and other commercial products left the impression that successful farming could no longer be performed without them. Moreover, the collapsed post-wartime period, during which this message crystallized within the land grant community, is nothing short of astounding.[41]

Arguments by CAC scientists about the essential role of biocides mirrored those being made by agri-state scientists throughout the United States during the 1940s and 1950s. While acknowledging the toxicity of organic chemicals and the problem of resistance, researchers also claimed that, when used as directed, biocides were safe and effective. Further, they consistently returned to the economic value of biocides in making more food and fiber available at lower costs to consumers. The testimony of Cornell entomologist Charles Palm during the Delaney Congressional hearings on pesticides in 1951 and 1952 was representative. Palm bemoaned the pervasiveness of DDT resistance but concluded that agricultural productivity depended on pesticides, which, in his estimation, provided "necessary benefits to public health and food production." During the same hearings, his colleague Stanley Freeborn, dean of the College of Agriculture at the University of California, Davis, argued that "wise and productive use of chemicals [was] of prime necessity and unavoidable" and "vital to our national economy." He further claimed that without modern machines and chemicals "our American agriculture would be out of business or reduced to peasantry." Freeborn's agri-state colleague Fred C. Bishopp, head of research at BEPQ, acknowledged the toxicity of petrochemicals, but he ultimately argued that they presented minimal risk and were of little consequence compared to their vast economic benefits. Four years later, USDA entomologist George Irving presented agro-apocalyptic predictions if chemicals were abandoned, claiming, "It has been estimated that insects take more production of our soils than man . . . if left uncontrolled, they would take at least half the average production of the U.S. acre. I'm not sure they wouldn't take it all." The overwhelming message emanating from land grant colleges and the USDA was that issues such as resistance and toxicity were minor and that embracing the new chemical paradigm was essential for the nation's health.[42]

Corn Transforms the Piedmont

After World War II, farms and feedlots applied the petrochemical imperative to a crop rapidly rising from marginality to dominance. During the 1930s, as Warren Monfort and W. D. Farr's new commercial cattle feeding business gained a toehold in Piedmont agriculture, a revolution in corn breeding brewed in the Midwest that would transform their businesses into feedlot empires. As an affordable and dense source of fats, starch, and digestible nutrients, no grain can compete with corn. The corn plants cultivated in the 1930s were adaptations of the maize planted by the indigenous peoples of North America and adopted by white settlers. Both utilized breeding techniques available to any knowledgeable farmer. Corn is a self-pollinating cultivar that depends on humans for its propagation. The male flower, which produces corn's pollen, exists in the tassel, while the female flower exists in the ear. During a sexual frenzy of five to eight days, tassels shed their pollen and fertilize the ears. Since the pollen from one plant will likely impregnate any number of plants—including itself—breeders cannot determine the lineage of any of the corn kernels they might plant the following season. Yet, corn's propagation depends on human intervention since corn kernels are locked in a husk that must be peeled before new seeds can be planted. Humans, for their part, can selectively breed corn for desired characteristics such as color, sturdiness, and tolerance to drought and climate by mating the pollen of one desirable plant to the ears of another, then saving the seeds for replanting. The process of breeding-in a single desired trait through this open pollination method can take more than a decade and involves much trial and error. Still, the skills of selective breeding are accessible to a dedicated farmer.[43]

Open pollination had economic and biological limits that were evident on the Piedmont and explain why Monfort and Farr imported their corn. Prior to 1940, corn was a marginal Piedmont crop for several reasons. Principally, the region's climate made the crop unpredictable. During some years, the season was long enough to bring a solid crop to harvest, while in others, late spring and early fall frost could kill off the entire crop. In addition, cool summer nights—colder than in the more humid Midwest—slowed corn's growth and retarded its yields.[44] Many dry-land farmers in eastern Colorado were willing to risk planting corn where yields averaged less than one-third of those achieved by Midwestern farmers. Low productivity could be justified since land costs were comparatively slight. Further, with little or no investment in irrigation, their choice of crops was limited by seasonal precipitation and soil moisture. By contrast, for Piedmont farmers with stock to feed, greater debt obligations, and the need to put irrigation to its greatest advantage, corn was risky. A poor year could impact both crops and animals. So, Piedmont farmers accepted the logic that sugar beets were the most lucrative crop to grow, and that its byproducts were a cost-conscious and efficient feed source for their animals.[45]

Even in Midwestern states, such as Iowa and Illinois, where corn thrived, open pollination had its limits. Despite improvements, average yields remained below thirty bushels per acre, a figure that hardly budged for fifty years. The breakthrough came as agri-state scientists and seed companies embraced principles of Mendelian genetics by investing in new strains of hybrid corn.[46] Hybrid production requires a much greater degree of breeding manipulation than was common among farmers. Breeders first developed inbred corn lines by selecting plants with desired characteristics and then allowing only self-pollination. By eliminating other plants from the gene pool over several generations, breeders could obtain genetic purity. These inbred plants possessed little value by themselves since purity resulted in weak yields. However, when researchers crossed two inbred lines, the resulting hybrid plants often outperformed commercially grown corn. The clear advantage of the new technology was that, given a critical mass of inbred plants with genetic purity, new breeds could emerge in one generation. Seed companies such as DeKalb and Pioneer first achieved commercial success with their hybrids in the Midwest during the mid-1930s, and by the end of World

War II, they dominated the nation's corn fields. Within ten years, yields increased by 50 percent, and by 1950, hybrids constituted 75 percent of the nation's corn acreage.[47] For the Farrs and Monforts of the world this was good news since higher yields generated a cheap and abundant supply of cattle feed.

Hybrid corn altered farmers' agroecological and consumer relationships. Hybrids do not reproduce their genetic characteristics from one generation to the next. Since their progeny consisted of several generations of inbred parents, hybrid vigor disappeared in the second generation as plants reverted to their weaker genetic inheritance. Consequently, it was useless for farmers to save seeds. Breeding hybrids required the skills of trained plant biologists and sufficient land devoted to breeding experiments and growing seed. So, farmers who wanted to tap the advantages of hybrid corn needed to purchase their seeds annually, essentially transforming seed producers into seed consumers. Moreover, seed companies patented each year's crop of hybrids, creating intellectual property out of genetic material in the public domain.[48]

Hybrid corn opened new Piedmont research frontiers as agronomists hoped that the new seeds could expand the Midwestern Corn Belt into the shadow of the Rockies. CAC quickly embraced corn hybrids and the seed companies that bred them. Between 1937 and 1939, the school's Fort Collins experiment station field tested sixty-five new hybrid corn varieties for seed companies. Pitting them against traditional open-pollinated varieties, they found that the hybrids outyielded their predecessors by 7.4–9 percent. Tests on newer varieties between 1942 and 1945 yielded ten to twenty more bushels per acre than the best open-pollinated breeds and produced 20 percent more silage. These field trials displayed a continuance of patterns first exhibited in seed treatments during the 1920s. Researchers at seed companies developed new seed strains annually. Experiment station scientists agreed to test them and provide feedback on germination, yield, and disease resistance. With limited fields on which to test the hybrids, CAC allowed each seed producer to contribute three seed varieties per experiment station, two of which had to be

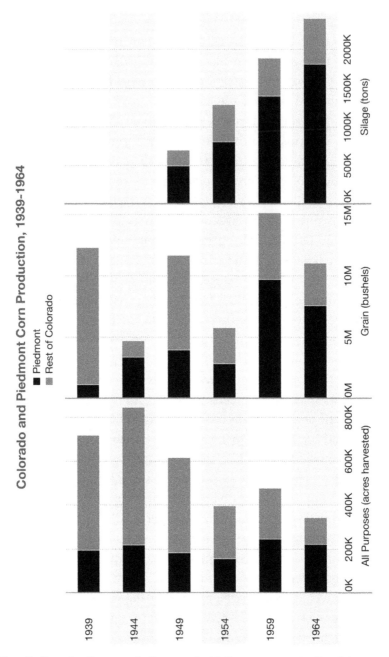

Fig. 28. Growth of corn and silage on the Piedmont, 1939–64. Graphic created by Philip Riggs and adapted from U.S. Department of Agriculture, *Colorado Agricultural Statistics* for the years 1939, 1944, 1949, 1954, 1959, and 1964.

ready for commercial production so that Colorado farmers could not only obtain test results but immediately purchase the most desirable seeds. Consequently, experiment station researchers functioned as product testers while offering a seedbed of competition for the new hybrids about to hit the market. As with seed treatments, hybrid testing offered another example of how agri-state efforts synchronized with industry prerogatives.[49]

Coupled with the emergence of commercial cattle feeding, the increased corn hybrid yields convinced irrigated farmers to plant corn. By the 1950s collaborative hybrid research and field trials conducted by seed companies and CAC produced yields well over one hundred bushels per acre—nearly triple what they were in the early 1930s, with no sign of plateauing.[50] Between 1940 and 1960, the seven counties that make up the Piedmont region of Colorado evolved from producing less than 10 percent of the state's corn grain to harvesting over two-thirds of it. Further, corn hybrids produced larger quantities of silage, a principal roughage in commercial cattle diets. Consequently, corn grown for silage in the region nearly tripled from 1949 to 1959.[51] The vigorous hybrids could be grown close together and possessed stalks that were straighter and sturdier than the varieties produced a generation earlier, enabling them to stand up to the rigors of machine harvesting. Moreover, since silage did not require mature ears of corn, it could be harvested earlier, reducing fears that an early frost could damage the crop.[52] Livestock consumed almost all of this corn. Consequently, while byproducts of the beet sugar industry remained a staple, they were quickly eclipsed by corn during the 1950s.[53]

Technology also played a part in corn's rise to Piedmont prominence. In the decade following World War II, local companies such as Harsh Hydraulics collaborated with manufacturers to develop machinery that could thresh and grind silage corn, depositing it into trucks equipped with hydraulic beds. Describing the process, W. D. Farr recalls driving these trucks into corn fields annually during the second week of September, processing silage, then dumping it into newly constructed pit solos where it would later be mixed with other feed rations and then delivered to waiting feed bunks. During the mid-1950s, Piedmont cattle consumed more corn silage than any other feed. By the foregoing method, Farr could supply a year's worth of silage for thirty thousand head of cattle. Farr's neighbor Richard Maxfield describes the impact of silage and machinery this way: "In the early days, we were shoveling everything . . . and then technology kicked in and we had a Harsh Hydraulic Truck, and that was one of the first ways we figured out how we can do this [easier]." Maxfield's family used their truck to haul corn grain from the local feed store as well. Consequently, he recalls, "We could buy and feed more cattle." In a little over a decade the marriage of seed biology and machine technology, energized by the collaborative efforts of company and college scientists, transformed a minor Piedmont crop into the driver of its most significant industry.[54]

Corn's rapid rise had dire consequences for Piedmont agroecology. Unlike sugar beets, which required careful attention to crop rotations and nutrient cycling, corn could be grown on the same ground year in and year out with heavy infusions of synthetic nitrogen and chemicals. Piedmont farmers like Richard Ulrich, whose family transitioned from sugar beets to corn after World War II, succinctly explained the tension between the two crops, "We just had too many beets on the ground for too many years . . . [but] you can raise corn on corn." How was this possible? The simple answer was that farmers such as Ulrich replaced the nitrogen formerly supplied by alfalfa and manure—which required crop rotations and patience—with the synthetic version that was available in critical quantities after the war. Statistical evidence draws a clear line between the Piedmont adoption of corn and its embrace of synthetic nitrogen. Between 1954 and 1959 commercial nitrogen sold in Colorado increased by 150 percent and its application on Piedmont corn accounted for approximately 40 percent of the total.[55] During the same period, organic sources of available nitrogen being applied to crops on the Piedmont declined for the first time in the twentieth century. Looking back at that decade, William R. Farr observed that corn "took over." He recalls that his family still applied feedlot manure, but that they applied "train-car loads" of synthetic nitrogen in the form of diammonium phosphate and ammonium nitrate to fertilize the ground in advance of planting corn. Then, during the growing season, they infused their irrigation ditches with anhydrous ammonia to resupply the crop.[56]

Monocropped corn also presented a weed problem begging for a chemical solution. Hybrid species bred for the Piedmont did not attract hordes of insects; however, their growth was particularly sapped by weeds.[57] Of course, when farmers poured irrigation water onto the new hybrids, weeds germinated with greater frequency and grew with vigor. Hand weeding was costly and inefficient, and the new hybrids were planted too close together for effective machine cultivation. Recalling efforts to monocrop corn without herbicides, Piedmont farmer Richard Seaworth stated that the crop would simply "fall down" within a few years. To make corn viable on irrigated land, CAC researchers told farmers to fumigate their soil in advance of planting with one of several compounds. At planting, they recommended using the selective herbicide 2,4-D that emerged from BEPQ experiments during World War II. It destroyed weeds that choked out corn plants without compromising yields. A substituted phenoxy compound, 2,4-D kills by forcing plant cells to grow uncontrollably, destroying vital tissues. According to CAC weed scientist B. J. Thornton, corn growers should apply 2,4-D every three to four weeks throughout the growing season to eliminate weeds. Summarizing the effect of these new biocides, Don Ament observed, "We never raised continuous corn before the herbicides."[58]

The high yields energized by hybrid seeds and synthetic chemicals and fertilizers were only possible when farmers and feeders were confident that irrigation water would be available in needed quantities. No single factor supplied that confidence more than the completion of the Colorado-Big Thompson Project (C-BT). Though not finished entirely until 1957, C-BT water flowed through every Piedmont canal by 1954. The C-BT increased the region's water supply by 20 percent and was more reliable than irrigation water originating in the South Platte watershed. In a drought, such as the one that wracked the region in 1954, water users absorbed its full impact since all of their water rights were invested in a single watershed that was fully tapped out even in the best years. By contrast, C-BT water originated at the headwaters of the Colorado River where water was abundant and Piedmont farmers' water rights were more secure. That plentiful water, along with the C-BT's early water right, created a virtual guarantee that the new water would be available on demand.[59]

The pronounced agroecological changes put in motion by the growth of the Piedmont corn-scape possessed a clear and pronounced correlation with C-BT water. Corn is both thirsty and finicky. High yields require more water and greater attention to precision irrigation than comparable yields of sugar beets. Since the C-BT provided an on-demand water source, farmers could be confident that the vagaries of the overworked South Platte Watershed could be smoothed out by the ready security of the Colorado River Watershed. That certitude not only moved corn from obscurity to prominence but encouraged the use of petrochemicals. Monocropping corn required fumigants to eliminate living organisms in the soil and abundant use of synthetic fertilizers to replace nutrients no longer supplied by nature. Farmers who grew corn on the same ground annually essentially replaced complex organic processes with simplified synthetic ones. Further, once conditioned by petrochemicals and fertilizers to grow corn, land cannot be flipped back to diverse crops on a timetable conducive to commodity farming. The combination of corn hybrids, biocides, and on-demand water implied a long-term commitment to a synthetic frontier. Recalling how the C-BT made corn a viable crop on the Piedmont town of Mead, Kent Peppler stated that, "prior to the C-BT, the main crop was sugar beets . . . and you might see a field of corn, but it was likely only for silage and there were not that many varieties." Fellow corn farmer Don Ament, whose irrigated farm still relies on return flows from the South Platte River, emphasized, "The Big Thompson Project made that river what it is." As if to punctuate the connection between corn and cattle, Ament continued, "I can't impress on you enough how . . . the whole cattle feeding industry [made] sure that [the C-BT] went through." Corn hybrids propelled the growth of commercial cattle feeding, and C-BT water placed corn abundance on the Piedmont landscape. The symbiotic relationship between the three of them unraveled many of the biological interdependencies that characterized the Piedmont prior to World War II.[60]

It's easy to see Piedmont farmers such as Kent Peppler and Don Ament as eithers pawns of agro-industry or economic rationalists. The former narrative suggests that they had no choice but to embrace the onslaught of biocides and hybrids to remain competitive in a rapidly industrializing agriculture. According to the latter argument, farming has always been about economic viability. Biocides, fertilizers, and cheap feeds promised to make lives easier and reduce labor costs while enabling farmers to manage more lands. This economic imperative suggests that the farmers of 1950 viewed their world through the same lens as those in 1870. After all, the Union Colony supported market farming and the technology needed to make it happen at its genesis. According to this logic, the only difference was in the technologies available to them. The Union colonists had soils to nourish, water to engineer, crop rotations to develop, and transportation networks to promote and tap into. Once their predecessors built that foundation, the Piedmont farmers of the 1950s could orient their lands around chemical, biological, and carbon energies unavailable eighty years earlier. Machines, biocides, and synthetic fertilizers, all driven by chemistry and cheap fossil fuels, supported rapid and relatively inexpensive yield gains.

One of the central flaws in this simplified historic logic is that it places environment and economy as polar and competing opposites. It is a story in which the demands of the free market always win out over the voices of nonhuman life. This rationale suggests that the W. D. Farrs and Warren Monforts of the world were either heroic entrepreneurs who led the farming community into a bold new economic future or environmental despoilers who ruptured an agroecological web that had been cultivated for over half a century. The reality of Piedmont change was more complex. Commercial cattle feeding on the Piedmont was not solely a function of entrepreneurial genius or depressed corn prices. The corn hybrids developed initially by agri-state and seed company scientists provided the cheap feed energy that accelerated the spread of commercialized animal feeding across the spectrum of species. Regional transformation was also driven by the distant labs of chemical companies who expanded their array of organic biocides and promoted them as solutions to organism inconveniences. Their efforts received a shot in the arm during World War II as the federal government enlisted those same companies to produce chemical panaceas for wartime diseases. Piedmont changes were also a consequence of conscious choices made by agri-state scientists to hitch their research wagons to agro-industry and to systematize Piedmont agriculture to fit their findings. Finally, the water generated by the C-BT made the possible probable for so many Piedmont farmers by convincing them that increased irrigation needs could be predictably met by an on-demand source. So, the eventual success of Warren Monfort, W. D. Farr, and those who followed them during the 1930s was more than a function of free market opportunism and a willingness to abandon a stable Piedmont agroecology. It was the result of contingent economic and political realities that functioned alongside and within a tangled set of public and private institutions who brought chemical and biological technologies to bear on the nation's farms and fields.

But what about decisions made by the majority of Piedmont farmers in the midst of the changes? Most operated in the broad space between commercial feeding innovator and agro-industrial pawn. The same farmers who had maximized the economic and feed utility of sugar beet agriculture and practiced nutrient cycling through systematized crop and livestock rotations were not idealists but practitioners of the possible. The economic, biological, and chemical transformations of the 1930s and 1940s revealed this reality. When an overabundance of farm commodities tanked crop prices, Piedmont farmers received less for what they grew, but they could import animal feeds cheaply. So, the commercial feeders who followed the Monforts and Farrs traded declining returns on sugar beets for potential gains, by running the cheap energies of corn through cattle, another depressed commodity. This is where economics, chemistry, and biology collided. The initial growth of commercial feeding rested on an assumption that corn's cheap availability would expand to fill the scale of industry growth. Until the early 1950s, the Midwestern Corn Belt had done exactly that. But when high-yield hybrids, chemical biocides, and abundant water all came together astride a burgeoning cattle feeding industry, Piedmont farmers jumped on board.

Those rapid changes reoriented the energies and complexities in nature. As Piedmont farmers abandoned the farmer/feeder model, they simplified organic processes. Crop rotations, raising livestock, employing manure and legumes to replenish the soil, and utilizing the by-products of sugar beet agriculture had formerly played critical roles in maintaining an agroecological sustainability based on complex interactions between organisms and minerals in the soil and air. Further, the rhythm of those interactions took time and patience. For example, nitrogen and phosphorous from manure and alfalfa were not immediately assimilable by crops. Instead, they required intricate chemical and microbiological reactions whose timing was neither immediate nor always predictable. Farmer/feeders planned for a complexity that existed where they lived, within the soils that they managed. While not operating in perfect balance, it was a model that supported economic imperatives and a diverse agroecology. By contrast, the farming and cattle feeding that replaced the existing model simplified the landscape. Managing soils could be accomplished by employing agro-industrial products to destroy unwanted organisms and then replacing them with synthetically generated ones. A sort of "kill and fill" mentality emerged whereby biocides destroyed the perceived enemies of crops and cattle, while farmers employed synthetic fertilizers, such as anhydrous ammonia, to add back to the soil what chemical agriculture had stripped. This could not help but upend intricate ecological interactions. The complexity that once operated in nature then moved away from the farm and into agro-industrial labs where researchers synthesized the latest compounds. While agri-state scientists still played a research role, increasingly they became product testers and demonstrators, translating a dense agro-scientific world for the consumers who would use its products.

The rapid-fire nature of changes to the Piedmont landscape during the 1940s and 1950s was impressive, though not total. During the early 1950s, the vast majority of cattle feeders still possessed fewer than five hundred animals, and most continued to cultivate crops. Sugar beet agriculture, despite an increasing dependence on biocides and fertilizers, still required crop rotation and soil management, and it remained an important driver of farming operations and the regional economy. These examples suggest that the Piedmont of the 1950s was transitioning into a hybrid of its former self, whereby farmers and feeders existed on a balanced spectrum between holding onto long-established practices and complete abandonment of them. But there were other forces at work on the Piedmont that pushed the hybrid world of the early 1950s to the fully articulated industrial feedlots of the 1970s. Those were driven by cultural, economic, and agri-state institutions, both within and beyond the region.

7 Manufacturing Beef

> Henry Ford would have felt at home in Greeley.
>
> —*TIME*, March 18, 1974

On a warm spring day in 1954, William R. Farr had an unenviable task. His father, commercial cattle feedlot operator W. D. Farr, instructed him to collect fresh manure—lots of it. An outbreak of bovine rhinotracheitis, also called red nose, was wreaking havoc on feedlot cattle throughout the Piedmont, threatening as much as one-quarter of the population. Red nose is one of several infectious bovine respiratory diseases (BRDs) that spreads rapidly in confinement. It causes delayed weight gain, pneumonia, aborted fetuses, and can be fatal.[1] Hoping to alleviate the red nose epidemic, the elder Farr cooperated with veterinarians and pathologists at Colorado State University (CSU) who diagnosed and treated sick animals.[2] They required fresh stool samples and the younger Farr served as data collector. Here is how he described it:

> I'd go into a pen of cattle and wait for one of them to do their deed and then go over to that little pile. I . . . collect[ed] samples from . . . about 20 to 25 animals. . . . Now I've got my jar full of shit to take over to the vet hospital at CSU and so I went over there just before they closed. . . . When I walked in, I tripped and this big gallon jug breaks on the floor and I've got . . . soft crap [all over the floor] and I don't know what to do with it. . . . I had to do the whole thing all over next week. This time [the veterinarians at CSU] gave me a cardboard container.[3]

Fig. 29. Farr Feedlots, 1946. City of Greeley Museums. Used by permission.

Farr's feces gathering expeditions paid off. Efforts by animal disease experts at CSU; pharmaceutical and chemical companies, such as Pfizer and Eli Lilly; the Colorado Cattle Feeders Association; and the yeoman work of stool scoopers such as William R. Farr led to the development of a partial vaccine within four years. Though it did not eradicate red nose, follow-up research showed that most cattle vaccinated prior to entering the feedlot environment developed immunity.[4]

When William R. Farr carried his stool samples to CSU, he tapped into a rapidly developing set of relationships between agri-state scientists, feedlot operators and their trade organizations, and a support network of agro-industrial companies supplying products to solve problems endemic to commercial cattle feeding. The disease vectors exacerbated by confining cattle required expertise not found in mixed farming, where crop rotation and seasonal livestock operations reduced bovine illnesses. A commitment to year-round cattle feeding, however, called for single-use confinement. Emerging pharmaceutical companies filled gaps with synthetic, disease-fighting antibiotics. This was not a direct pharmacy-to-farmer relationship. As in the past, industry sought agri-state cooperation. By the 1960s CSU, like many schools within the land grant college complex, intensified its relationship with agro-industry, institutionalizing its connections by hiring research faculty vetted by cattle feeders.[5] The web of networks that originated in Farr's pens, saturated the Piedmont, and radiated

outward to far-flung labs and factories illustrates how commercial cattle feeding both reflected and articulated relationships that had been developing for over half a century.

These intensified relationships enabled cattle feeders to reframe how farmers perceived their livestock and the nature of their lifecycles. When leading cattle feeders coalesced to court agro-industry and land grant scientists, they achieved an insider status and a platform to influence their peers. They became sought-after speakers, influenced the hiring of university faculty, directed research agendas, and were consulted at various levels for their expertise. Successful feeders were also admired among their peers. According to Kent Peppler, who grew up farming on the Piedmont during the 1960s and 1970s, elite cattle feeders were "like movie stars to us."[6] Their message emphasized economies of scale that, directly or indirectly, squeezed out mixed farmers and small operators. Commercial feeders argued for a regimented, corn-based diet, alongside synthetic supplements, in a single-minded focus to reduce feed costs and compress cattle lifecycles.[7] These operators also worked to calibrate feeding and meat processing with perceived consumer demands and specific USDA standards. All of this functioned to narrow feeding operations and constrict the diversity of agriculture that could support it. By 1970 cattle feeding on the Northern Colorado Piedmont mimicked manufacturing operations—in form and function—with little space for the sustainable farming model out of which it evolved.

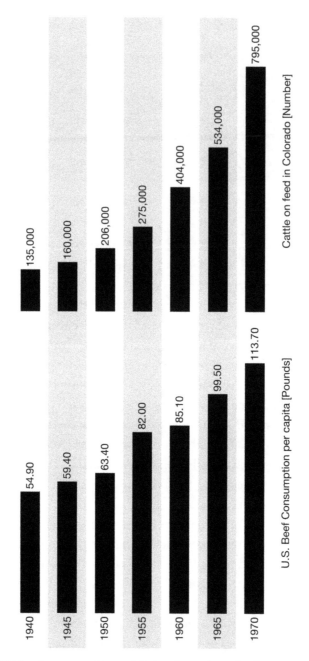

Fig. 30. U.S. beef consumption and Colorado cattle on feed, 1940–70. Graphic created by Philip Riggs and adapted from materials in the W. D. Farr Archives, CSU Morgan Library Archives and Special Collections.

Ultimately, the cattle feeding complex overwhelmed the farmer/feeder model that preceded it, reordering the landscape. As quantities of Piedmont cattle grew exponentially to meet demands for cheap beef, so did industry demand for corn grain and silage. With yields increasing due to hybrids engineered for Piedmont landscapes and the ability to monocrop year after year using petrochemicals, corn was an attractive proposition. To meet their insatiable demands, feedlot operators courted advanced contracts with local farmers to supply their needs, giving farmers guaranteed buyers. At the same time, powerful herbicides reduced labor needs and decimated soil biodiversity, rendering potent soils inert for years. In the background, increased water, primarily from the Colorado-Big Thompson Project, coursed through Piedmont arteries. Cattle and corn became a repository for all of these biological, chemical, and hydrological transformations, and feedlots were the physical location where all of these technologies converged. As former Colorado agricultural commissioner and lifelong Piedmont farmer Don Ament stated, "All of that corn goes through those cows. . . . Everything, in fact, goes through those cattle."[8]

Marketing the Pharmaceutical Revolution

During the early 1950s, as corn hybrids gained a foothold in Piedmont agriculture and petrochemicals became commonplace, another synthetic revolution brewed in Piedmont cattle bodies. As feeders embraced DDT and other chlorinated hydrocarbons to eradicate pests that targeted cattle from the air, agri-state researchers and leaders in the pharmaceutical industry tested antibiotics and hormones that could simultaneously kill disease pathogens and accelerate the timetable to slaughter. Cattle arrived at feedlots traumatized. After being raised on pastures for the first four to six months of life, feeders purchased them and moved them by truck to a feedlot.[9] Cattle lost weight in transport and arrived dehydrated, exhausted, and with little appetite. The traumatic journey from pasture to feedlot also exposed animals to multiple diseases, the most common of which were BRDs, commonly called shipping fever, since they were typically contracted in transit. At best, BRDs delayed feeding. At worst, they resulted in severe pneumonia and death. Once feeders confined cattle into pens and placed them on full feed rations, the race to gain enough weight for slaughter possessed further medical pitfalls. Since cattle had not evolved to eat corn grain and silage, the principal rations, it was common to contract liver abscesses. These resulted in condemnation of the animal upon slaughter.[10]

Even if animals remained disease free in confinement, there were limits to how quickly they could convert grains and roughages into meat. Though research during the first half of the twentieth century had enabled cattle to fatten 42 percent faster on 30 percent less feed, most of the possible gains from adjusting rations had been achieved.[11] In the postwar period, the frontiers in concentrated animal feeding operations—cattle, poultry, and swine—were found not in achieving the right balance of feeds but in manipulating biological factors to make existing animal diets reshape natural processes, while at the same time eradicating the disease vectors endemic to confined feeding operations. Opportunities to expand biological boundaries attracted private laboratories and pharmaceutical companies to the milieu of research conducted in feedlots and at agricultural colleges.[12]

Antibiotics were the first feed additives that received significant attention by the feeding industry. According to historian Terry Summons, the lack of meat protein available for human consumption during World War II combined with fewer available animal feeds stimulated industry queries on how existing rations could translate into more meat. Researchers at American Cyanamid, a leading chemical and pharmaceutical corporation, demonstrated that penicillin and tetracycline could speed up growth in chickens by 10 percent. Eli Lilly and Merck, both growing pharmaceutical companies, found that antibiotics synthesized from vitamin B12 could improve feed conversion ratios by a modest 3–4 percent in cattle while minimizing the bloat common from eating high quantities of grain. By 1951 the FDA had approved several antibiotics for use as feed additives and for disease prevention, including penicillin, terramycin, and aureomycin. According to Purdue animal scientist William Beeson, in a speech made to the National Cattleman's Association in 1959, combinations of these feed additives, fed at 70–80 mg/day, increased daily weight gain by 10 percent, feed efficiency by 8 percent, and improved the overall quality of the carcass, leading to higher purchase prices. At a cost of less than a penny per head each day, it was among the most effective methods for increasing profit. Statistics provide evidence that feedlot operators paid attention. In 1951 farmers spent $17.5 million on antibiotics for their animals. Ten years later, that figure nearly tripled.[13]

Concurrent with efforts to find a red nose vaccine, W. D. Farr employed his cattle in another experiment involving land grant researchers and feedlot drugs. Following the lead of colleagues investigating the use of drugs in chickens and pigs, Iowa State animal scientist Wise Burroughs began feeding and implanting cattle with the synthesized female growth hormone diethylstilbestrol (DES), hoping that it would increase feed efficiency. Noting some initial success, Burroughs sought out a critical mass of cattle to field test DES. Farr was an enthusiastic participant. So, for three consecutive summers in the early 1950s, animal science graduate students from Iowa State descended on Farr's feedlots to collect data from animals fed and implanted with DES. The 1953 experiments were the most successful. According to Burroughs, they showed that fed at 5 milligrams daily, DES induced cattle to gain weight 35 percent faster and reduced feed costs by 20 percent. Follow-up studies showed more modest gains of 10–20 percent. Anticipating FDA approval, the pharmaceutical company Eli Lilly signed a contract with Iowa State to manufacture the growth hormone. Despite concerns over whether DES residues remained in animal tissues at slaughter and could be cancer causing in humans, the FDA approved DES in November 1954. Sensing a gold mine, Eli Lilly began manufacturing Stilbosol as a feed additive four weeks later. Concurrently, Eli Lilly used DES as the platform to launch its new animal pharmaceuticals division, Elanco. The following month, January 1956, CSU began field trials with DES at its experiment station in Fort Collins. The impacts were electric. In less than one year, 50 percent of all cattle raised in the United States were fed DES orally, while W. D. Farr claimed that 80 percent of all Piedmont feeders employed the hormone. Farr took further advantage of his

Fig. 31. W. D. Farr advertises Stilbosol. IUPUI University Library Special Collections and Archives. Used by permission.

position as DES pioneer by mixing it with feeds and selling them from his Greeley feedlot. By the early 1960s, 95 percent of the nation's feedlot cattle ingested DES as part of their normal feed regimens. It was easy to understand why. According to DES-feeding surveys conducted in 1959, the typical feeder employing DES returned $11.50 for every dollar spent on the hormone. DES offered a biological cash cow for commercial feeders.[14]

While the practices and promotion of feeders such as W. D. Farr provided incentives for peers to employ drugs in feeding regimens, support for widespread adoption also came from one of his professional organizations, the Colorado Cattle Feeders Association (CCFA). Founded by Farr and his commercial feeding peers in 1955, at a moment when red nose ravaged Piedmont cattle and DES was poised to revolutionize the industry, CCFA encouraged increased beef consumption, lobbied politicians at the state and national levels, and published a monthly magazine for its members called *Over the Feed Bunk*. In addition, CCFA entered into an agreement with CSU in 1955 whereby elected members drawn from the association and the college oversaw the cattle feeding "research projects, facilities, and personnel" at CSU. This gave CCFA remarkable influence. While it was no secret that agricultural research at CSU often took its cues from agro-industry, this agreement formalized that relationship and gave it contractual standing. Feeders paid an annual membership fee based on the number of cattle they fed, while auxiliary members, including Piedmont veterinarians and pharmaceutical companies, joined to wield influence and expand clientele.[15]

During its early years CCFA's most popular initiative was its feedlot drugs program. Beginning in 1957, CCFA contracted with pharmaceutical companies to purchase drugs at discounts that were then passed on to members. Within a couple of months, CCFA sold more than $16,000 in feedlot drugs.[16] While many of the drugs were purchased by feedlot operators at CCFA offices or through the mail, feeders often found it more convenient to buy them through member veterinarians. The program enabled CCFA to expand its network of influence. Manufacturers of feedlot pharmaceuticals such as Merck,

Eli Lilly, and Pfizer joined CCFA, advertising their products in association literature, hoping to boost sales in the process.[17] Feedlot drugs quickly became a conduit through which commercial cattle feeding, the pharmaceutical industry, and CSU intensified their relationship and clarified shared agendas. With startling rapidity, synthetic hormones and antibiotics went from designer novelty to universally accepted good within the industry.

The specific drugs purchased from CCFA suggest that feedlot operators were concerned with diseases and parasites that were endemic to commercial cattle feeding and could hinder the ability of cattle to quickly convert feed to flesh. Consequently, pharmaceutical companies, including Eli Lilly and Cutter Labs, offered Piedmont operators antibiotics such as bacitracin, Boceptol, and sulfamethazine, valued principally for their role in accelerating the shift to full-time feeding. Once cattle recovered from the shock of their forced relocation, feeders concerned themselves with the liver abscesses and bloat associated with heavy corn-based feeds. For these ailments, CCFA purchased and sold antibiotics such as tylosin, manufactured by CCFA member Eli Lilly. CCFA also sold drugs aimed at controlling parasites, such as cattle grubs, and various flies that swarmed to feeding pens. Mono-cropping cattle profitably required that these factory-fed animals run the gauntlet of diseases associated with a life and diet foreign to their biology. Partnerships between pharmaceutical makers and CCFA played a pivotal role in the network of Piedmont relationships making that possible.[18]

The animal scientists and veterinarians at CSU were indispensable parts of the support infrastructure for the adoption of feedlot drugs, eventually developing a robust program of their own. Following World War II, CSU trained veterinarians specifically to work with feedlot cattle, furnishing the majority of those working in the Piedmont feeding industry. While the largest operators employed full-time veterinarians on-site, most routinely hired CSU vets to vaccinate their cattle and treat specific diseases such as BRD and dystocia.[19] In addition to the work of veterinarians, researchers at the CSU Experiment Station, taking some of their direction from CCFA, conducted follow-up DES studies beginning in 1955. Partnering with eight land grant colleges, results showed that DES enabled cattle to gain weight 16 percent faster on 13 percent less feed when compared to cattle who did not receive the hormone. CSU researchers also conducted studies on the use of higher levels of DES to induce abortions in cattle.[20]

The network of relationships promoting the use of pharmaceutical drugs and other additives in cattle feeding rations extended to small feeders and regional retailers as well. James Svedman, an active member of CCFA, farmed various crops and operated a commercial feedlot near Fort Collins from the 1950s through the 1970s. Svedman typically fed around one thousand steers and heifers at a time. While he purchased some drugs through CCFA member veterinarians to address specific diseases, he spent much more on feed additives containing hormones and antibiotics. In fact, just a few years after CCFA began selling feedlot drugs at reduced prices, local retailers and feed mill operators incorporated synthetics into their feed supplements. For its part, CCFA was glad to concede this role since proliferation, not profit, was its principal aim. In the late 1950s, Svedman made regular purchases of "Feedlot Fattener #1," manufactured by Ranch-Way Feeds and Sunland Feed, both located in Denver. Fed at two pounds per animal per day, the pellets provided ten milligrams of DES to each steer or heifer. By 1963 the fattener had become a staple within Svedman's feeding arsenal and it came fortified with Stilbosol and seventy milligrams of aureomycin.[21] Larger operators could avoid retailers altogether by purchasing drugs direct from pharmaceutical companies and mixing them with other feeds and additives using their own mill. W. D. Farr went one step further, employing his feed mill not only for his own cattle but also to concoct made-to-order supplements for his peers, cleverly naming his business Farr Better Feeds.[22]

Pharmaceuticals shaped the thinking of feedlot operators on the Piedmont in ways that their petrochemical cousins did not. Synthetic biocides and fertilizers, though products of labs and manufacturing complexes rather than nutrient cycling and historic farm praxis, still filled fundamental needs such as eliminating insects and weeds and replenishing nutrients lost by crop cultivation. Available nitrogen, whether the result of complex microbiological reactions or synthesized with the help of fossil fuels, remained essential to plant growth. Pharmaceuticals, by contrast, did not replace something lost but added something heretofore unthought of or unneeded. The feeding efficiency results of the pharmaceutical revolution in agriculture, unlike the chemical one, could not be replicated through good farming practice, since the new feed additives claimed to do more than address a pervasive agricultural problem. By speeding up biological processes, synthetic drugs and hormones solved problems that were previously nonexistent using products whose function had little or no precedent. Pharmaceuticals did not just mimic natural processes. They altered nature's clock.

Factories in the Feedlots

No feedlot operator on the Piedmont had a greater influence in moving peers toward embracing new practices than W. D. Farr. By the mid-1950s, Farr owned several farms near Greeley and a commercial feedlot that handled eight thousand cattle. Farr bought and sold cattle throughout the year, employed a full-time nutritionist, and had diversified his operations to include selling feed mixes and growing potatoes to sell to chip companies. Ten years later, Farr had computerized the mixing of cattle feed and deployed hydraulic machines to deliver them. Farr's work with the pharmaceutical industry, as well as with CSU and other land grant colleges, gave him a recognizable name within commercial feeding circles. He also possessed a folksy charm and natural charisma that drew audiences.[23]

During the 1950s and 1960s, Farr employed his influence at forums frequented by cattle feeders, allied researchers, and agro-industry. He also attracted peers eager to view his Greeley operations.[24] Farr's message was simple, straightforward, and rarely waivered. He told cattle feeders that they should replace any sentimental notion that they were farmers or engaged in husbandry. Rather, in the words of a 1957 speech, they were engaged in a "manufacturing operation geared to convert feed into food in the most efficient manner . . . with emphasis on the kind of beef that the consumer wants."[25] In order to do this, Farr stated, feeders should purchase cattle throughout the year and maintain the greatest quantity of feed pens they could manage, in order to maximize their market advantages. He told his audience to pay little attention to popular contests that judged the size and attractiveness of cattle. To Farr such competitions distracted feeders from more important metrics such as feed-conversion ratio and obtaining corn grain and silage with the correct moisture content at the lowest price.[26] Dismissing lingering concerns about the need for cattle to graze on grass for significant periods before they entered the feedlot, he told his listeners to buy young and light cattle and to place them in pens quickly to take advantage of periods when animals grew at the fastest rate. Finally, when it came to scientific advances in the field, Farr played the twin roles of prophet and promoter, at once reminding them of the value of hormones, antibiotics, tranquilizers, and feed additives while alerting them to pay attention to new frontiers in the field. Supplanting animal biology, Farr envisioned a future where every desirable feature in beef producing cattle could be bred into one bull, the semen of which could be universally used in all feedlot operations. As a result, according to Farr, "Every calf would produce tender beef . . . have the maximum amount of high-priced cuts . . . [possess] the characteristics that produce high-yielding carcasses, [and] produce twice the daily weight gain we now get." For Farr, feedlots, and the animals residing within, were factories calibrated to employ science, consumer behavior, and marketing savvy to manufacture a profitable product: beef.[27]

The transformations W. D. Farr referred to in his speeches and interactions and practiced in his own feeding operations, suggested radical changes in Piedmont agroecology on the horizon. The type of operations Farr described were incompatible with the mixed husbandry that small farmers practiced. Maintaining multiple pens, keeping abreast of market trends, constantly buying and selling cattle, and purchasing a host of feeds and supplements in varying quantities did not synchronize with the mixed husbandry and nutrient cycling that predominated prior to World War II. The former emphasized sufficient capital and mastery of the built agricultural landscape, and it viewed science as a means to supersede slow-moving ecological processes. The latter required attention to the interaction of existing organisms and viewed chemical interactions primarily as a set of processes occurring within the ecology of a farm. Most farmers now had to choose between factory feedlots and crop farming. In fact, Farr stated as much in 1957, when he told a group of Iowans, "Farm units must get bigger to support the mechanical age we are living in."[28]

For Farr, it was both inevitable and desirable that specialization and size replace small farmers practicing mixed husbandry. Further, he affirmed how confined cattle feeding operations were changing human/animal relations. With a laser-like focus on feeding efficiencies and consumer demand, Farr convinced his peers that cattle were meat factories. Faster weight gains and younger cattle meant that lifecycles were shortened to the point where the overall nutrition and health of the individual animal could be disregarded as long as it could survive to slaughter. And what about all of that manure? With less land devoted to crops and a greater volume of feces generated, the manure that once generated life became waste. If W. D. Farr's assessment was correct, then cattle feeding remade agriculture in the postwar era into a manufacturing operation.

Feeding Beef: John Matsushima

CSU animal sciences professor John Matsushima translated W. D. Farr's manufacturing ideal into a research agenda that was financed and directed by a broad cast of agro-industrial actors. A third-generation Japanese American, Matsushima was born in 1920 to parents who worked in Piedmont sugar beet fields. He grew up surrounded by German Russians whose families made the transition from beet laborers to growers for Great Western Sugar. In his early years, Matsushima witnessed the mixed farming that emphasized crop rotations, seasonal cattle feedlots, and nutrient cycling. While an active member of both 4-H and Future Farmers of America, he developed a knack for feeding steers, competing with members of the Monfort family at stock shows, occasionally besting these pioneers

Fig. 32. Photograph of John Matsushima, 1961. University Historic Photograph Collection, CSU Morgan Library Archives and Special Collections. Used by permission.

of the commercial feeding industry. In 1939 Matsushima earned a National Merit Scholarship to attend CSU where he majored in animal husbandry, receiving both a bachelor's and master's degree. He then completed a doctoral program in animal science at the University of Minnesota, where his dissertation research compared the "fattening performance" of multiple cattle breeds. Matsushima then took a position in the Animal Sciences Department at the University of Nebraska, where he remained through the 1950s.[29]

The circumstances surrounding Matsushima's return to Northern Colorado offer insight into the developing networks between the feeding industry and agricultural colleges.[30] Because the commercial cattle feeding industry was relatively young in 1960, word of John Matsushima's expertise traveled rapidly, making him a desirable commodity and prompting efforts by feeders to lure him to their states. Though Matsushima received inducements from feeders in Kansas, Oklahoma, and Texas, he elected to take a position in the animal husbandry department at CSU in 1961. Led by Warren Monfort, CCFA told Matsushima that they were uniformly in support of finding a position for him at CSU. They argued that he would have no shortage of opportunities to carry out his research agenda, that CSU was invested in cattle feeding research, and that he would find a bevy of dedicated graduate students eager to work with him. As Matsushima explained, he initially balked due to moving costs and his commitment to graduate students at Nebraska. Undeterred, CCFA offered to pay Matsushima's moving expenses and bring both graduate students and the cattle that were his research subjects from Nebraska to CSU. As Matsushima implied in multiple oral histories, once he succumbed to industry enticements, interviewing and signing a contract with CSU was a minor matter. In fact, Matsushima recalls the presence of several prominent feeders during his university interview. As Matsushima's recollections suggest, CSU had already taken the junior role in its partnership with CCFA.[31]

Matsushima's hiring at the behest of industry was not uncommon during the period. Ray Chamberlain, who was dean of CSU faculty in 1961, eventually becoming university president in 1969, explained that hiring Matsushima was part of the larger trend among agricultural colleges to become well-respected research institutions. New faculty were courted with a view to their ability to attract industry research dollars. Since Matsushima was already well connected with the cattle feeding industry in 1961, his hiring was as much a validation of the research money he could attract as it was for his skill. In fact, as Chamberlain implied, the university deemed industry funding and research prowess to be synonymous. Further, as former CSU weed scientist Robert Zimdahl explained, universities such as CSU discouraged research independence during the period, as salaries and research funding were kept low as part of efforts to push faculty to seek out industry money. The fact that Matsushima arrived with a hearty endorsement from feedlot operators validated both his research acumen and his fund-raising savvy.[32]

The university's emphasis on courting industry support and acceding to its wishes played out through Matsushima's early career at CSU. Initially, much of his time was occupied with assessing the needs of cattle feeders and changing the culture of feeding to match his findings. Cooperating with extension specialist James Sprague, Matsushima assessed industry needs by visiting 460 commercial feeders in Colorado.[33] One concern uttered by the larger feeders, including W. D. Farr and Warren Monfort, was that cattle culture wrongly emphasized large, physically appealing cattle. At events such as the annual National Western Stock Show in Denver, the highest prizes went to these animals. Matsushima shared this assessment. During his period in graduate school and his time in Nebraska, Matsushima had many opportunities to disassemble cattle carcasses. He concluded that cattle should be slaughtered at weights lighter than industry norms. But light cattle were not prized by judges. The goal of feeding cattle, according to Matsushima, was to produce the beef that consumers craved. Moreover, the principal objective of livestock judging was to reward feeders whose operations were driven by market demands. Repelled by the prevailing cattle culture, Matsushima quit judging livestock in the 1950s after an event in Omaha in which the grand champion steer "had the worst carcass," with "an inch of fat" all around it. Consequently, when Willard Simms, the superintendent of the National Western Stock Show approached Matsushima about developing a new prize category for cattle, he saw the opportunity to tilt the culture of cattle raising in favor of commercial feeders. Matsushima named the new category the "Fed Beef Contest."[34]

Starting in 1965, judges at the Fed Beef Contest awarded prizes for cattle based on health and USDA grading. Cattle received lower scores if they showed any signs of pest infestation, such as attacks from cattle grubs, since this was an indicator of animal stress that could reduce meat quality. They were also judged on yield grade—also called cutability—which was a measure of the percent of retail yield for the chuck, rib, round, and loin portions of the beef. Finally, animals were evaluated on whether they attained the USDA Choice grade. While only the second highest beef grade (Prime was the highest), Choice was desired by consumers due to its combination of affordability and taste. Assessing grades occurred post-slaughter, when judges evaluated the intramuscular fat—also called marbling—in the rib-eye muscle between the twelfth and thirteenth rib, and the age of the carcass. While fat around the edges of beef was undesirable to consumers, evaluators regarded thin layers of marbling within each cut of meat as markers of flavor. Younger carcasses yielded higher grades than older ones since meat tenderness decreases with age. In its initial years, the Fed Beef Contest also employed a category called "confirmation," which measured an animal's overall appearance. Here, Matsushima influenced changes again. After two animals whom he scrutinized as too fatty won due to high scores in the confirmation category, Matsushima argued that the criteria should be eliminated. Stock show organizers gladly complied.[35]

In addition to normalizing the USDA beef grading system, the Fed Beef Contest was notable for the nature of its impact on the culture of raising cattle. Once the confirmation category was eliminated, winning was entirely determined by post-slaughter analysis and encouraged commercial feeding practices. The contest devalued personal relationships with animals since pride in appearance only mattered insofar as an animal was free of disease. That goal was largely accomplished through assembly-line application of vaccines and biocides, along with factory-mixed supplements laced with antibiotics. Further, Matsushima and his peers understood that corn-fed cattle attained a higher degree of marbling than those finished on grass.[36] This benefited commercial feeders who knew that a quick transition from grass to corn as the primary feed propelled faster weight gains and led to slaughter at younger ages, thus increasing the likelihood of higher grades. Matsushima argued that the Fed Beef Contest and similar competitions at other stock shows were instrumental in the growth of commercial feeding and in aligning consumer demand with industry practices.[37]

Chuck Sylvester, former executive director of the National Western Stock Show, agreed with Matsushima. During the 1960s, the stock show was a critical nexus for understanding the status and direction of the beef cattle industry. As the largest stock show in the United States, it attracted cattle feeders, representatives of agro-industry, farmers of every stripe and age, students and faculty of animal sciences at CSU, and a curious meat-eating public looking for some entertainment. Chuck Sylvester was one such patron. A fifth-generation Piedmont farmer living on a family homestead settled in 1866, Sylvester grew up in the 1940s and 1950s when "every farm had a small feedlot and fed out cattle." Sylvester was enthralled by the stock show. As a member of both 4-H and Future Farmers of America, he recalls preparing animals for judging. In 1967, while working on an animal science degree at CSU, Sylvester became the superintendent of livestock judging at the stock show, giving him an opportunity to work closely with both Willard Simms and John Matsushima.[38]

During the course of the Fed Beef Contest, the public could view live cattle in one building and then walk across the street to buildings once occupied by meatpacking giants Swift and Armour, where they could view the carcasses of those same animals immediately after slaughter. Sylvester recalls kids being "so shocked when they watched the process because they had never been in a packinghouse." Patrons could then view animals being graded and butchered before their very eyes, then speculate on awards before listening to judges' pronouncements. Once winners were announced, officials placed top carcasses on display, using them as tools to educate the public on the processes by which they obtained the most flavorful steak in the market. To cattle feeders, agri-state scientists, and agro-industry, the Fed Beef Contest collapsed the time and space between themselves and the beef eaters they targeted. It gave faces to consumer abstractions. The Fed Beef Contest also provided an education to cattle feeders and agri-state scientists. According to Bill Hamerich, former president of CCFA, his cattle-feeding peers studied the results of the Fed Beef Contest. They wanted to know how the winners were fed, what breeds scored highly, and who sired them. Don Ament recalls that when his grandfather tried to improve his herd, he would buy bulls whose sires were contest winners. The Fed Beef Contest was both a microcosm of the cultural and economic transformations in cattle feeding and an important driver of the changes it displayed.[39]

The Fed Beef Contest was also a place where the work of John Matsushima, the agri-state scientist, and W. D. Farr, the feedlot operator, came together. While Farr, a regular participant in the Fed Beef Contest, convinced his peers to view their feedlots as factories calibrated to construct beef efficiently, Matsushima's disassembly of

Sampling of John Matsushima's Industry-Sponsored Field Trials from 1974–1976

Study	Funding	Dollar amount
Feed Truck	Hydraulics Unlimited (Harsh)	$16,038.00
Pro-Sil	Ruminant Nitrogen Prod.	$11,250.00
Stirofos	Shell Development Co.	$27,425.00
Niacin	Lonza, Inc.	$10,000.00
Monensin & Liver Abscess	Eli Lilly & Co.	$5,000.00
Fellowship	Ralston Purina	$4,000.00
NPN Utilization	Moorman Manufacturing Co.	$4,000.00
Tyloxin	Eli Lilly & Co.	$4,000.00
Hormone	Myzon Laboratories	$3,000.00
Famphur	American Cyanamid	$3,000.00
Animal Waste	Ceres Land Co.	$1,000.00
NPN Utilization	Mitsubishi International Inc.	$14,500.00

Fig. 33. Sampling of John Matsushima's industry-sponsored field trials, 1974–76. Graphic created by Philip Riggs and adapted from Colorado Cattleman's Association, CSU Morgan Library Archives and Special Collections. Used by permission.

cattle carcasses positioned him to reorient the modern stock show around industrial values. Their place at the stock show, however, suggests that the cultural and economic changes in commercial cattle feeding operations could be viewed from both ends of the bovine lifecycle. Farr focused his efforts on engineering cattle. Feeds, pens, chemicals, pharmaceuticals, and consumer awareness were all calibrated to construct the beef-producing technologies that fed cattle. By contrast, Matsushima deconstructed cattle to change the industry. In stripping hides and removing life-giving organs from steers and heifers and then placing the resulting consumer product on display, Matsushima synchronized the actions of the cattle-feeding complex with the desires of the beef-eating public.

John Matsushima also conducted wide-ranging research on the mechanics of cattle feeding. One of the significant costs for commercial operators was mixing the increasing number of feed additives into cattle rations. During the early 1960s, Piedmont-based Harsh Hydraulics invented a mobile product modeled after a cement mixer. Capitalizing on the idea, Matsushima and his graduate students in CSU's Beef Cattle Research Center developed a tractor-driven version in 1965 capable of continuously mixing feed and distributing it to feedlot pens. Ten years later, Matsushima and his graduate students mounted a Harsh Hydraulics mixer onto a commercial grade truck, vastly increasing the quantity and versatility of mixing and transporting feed. As the science of manufacturing beef incorporated a complex bevy of antibiotics, hormones, tranquilizers, vitamins, and minerals into each bovine meal, Matsushima's inventions calibrated and transported the rations.[40]

Harsh Hydraulics offers a further window into how science, land grant institutions, and industry mixed to serve the modern feedlot. The Harsh feed mixer was just one among dozens of projects Matsushima and his students worked on as part of the CSU Department of Animal Studies' Beef Nutrition Project. On two separate parcels of land owned by CSU, they operated a combined feeding and research operation. As with feedlot operators throughout the Piedmont, Matsushima and his students purchased and sold cattle at market rates.

They concocted, mixed, and delivered feeds while conducting a host of experiments. Though the Beef Nutrition Project received funds from CSU, the university expected it to be self-sustaining.[41] And it was no small operation. Between 1968 and 1977, the project bought and sold between $500,000 and $1.2 million in cattle annually, frequently feeding more than two thousand animals at a time. While not operating on the same scale as the Monforts or the Farrs, these figures placed the CSU operation in the "commercial feeder" category.[42] Measuring the economics of feeding another way, Matsushima jokingly stated that the Beef Nutrition Project "brought in more money than the budget of the entire [CSU] athletic department."[43]

While intended to mimic free market commercial feeding operations, the primary value of the Beef Nutrition Project's size was that Matsushima and his students could scale research projects to meet industry needs with the aid of industry dollars. This is where the interwoven relationships between the modern land grant university and agro-industry are most visible. "Feedlot Performance" was one of the research categories supervised by Matsushima. Its goals were to increase feed-conversion ratios, lower production costs, and increase the quantity of cattle that would grade USDA Choice. While research varied year over year, funding and agendas were largely shaped by feedlot operators, meatpackers, and manufacturers of feed products. The project's budget for the fiscal year 1974 to 1975 is illustrative. While the university awarded Matsushima slightly more than $31,000 to conduct research, he received approximately $129,000 from agro-industry.[44]

Most projects involved testing a new feed additive, meeting environmental regulations, or addressing diseases endemic to commercial feeding. For example, Monfort appropriated $7,500 in the hopes that it could turn cattle waste into cattle feed, since new Environmental Protection Agency regulations in 1972 curtailed the ability of feedlots to dump wastes into local waterways.[45] By 1971 Monfort was not only fattening cattle but slaughtering and processing its own beef. One of the waste products of slaughter was undigested feed, called paunch material. Monfort was one among many slaughtering operations who hoped that undigested food could be dried and then incorporated into feedlot meals. This research project empowered Matsushima and his team to employ the stomachs of unsuspecting cattle for restarting digestive processes that had initially been cut short by the industrial timelines of the modern feedlot. Shell Corporation was the single largest contributor to Matsushima's budget that year, giving $26,500 to perform field trials using stirofos, an organophosphate intended to kill grubs, flies, and other insects that fed on cattle. Eli Lilly, one of the United States' largest pharmaceutical companies, provided money for four separate projects. Two of them involved the use of drugs to reduce diseases exacerbated by the particularities of commercial feeding. Eli Lilly hoped that Tyloxin would noticeably decrease the occurrence of bovine respiratory disease. Eli Lilly also granted Matsushima funds to experiment with a new drug called Rumensin, intended to reduce coccidiosis, a disease caused by cattle eating their own feces, common in feed pens. Concurrently, Matsushima tested Rumensin's ability to reduce the liver abscesses caused by the grain-based diet in feedlots; another example of employing pharmaceuticals to compensate for diets nature did not intend. Among other projects, Matsushima received grants for synthesizing proteins in cattle rumens, subtherapeutic uses of antibiotics, and new synthetic hormones to replace the recently banned diethylstilbestrol.[46]

John Matsushima's work illustrates how completely agri-state science had assumed both the culture and the research agenda of agro-industry. From his time as a graduate student evaluating the fat content of recently slaughtered cattle to his long career at CSU, Matsushima unwaveringly pursued projects aimed at efficient feeding, reducing operator costs, curbing endemic cattle feeding ailments, enabling chemical and pharmaceutical companies to bring products to market, and aligning beef consumption with industry practice. Moreover, industry leaders wooed Matsushima in the hopes that his work would further tilt the culture of cattle feeding in favor of commercial feeders. By courting Matsushima to develop the Fed Beef Contest and inviting him to play a prominent role in livestock shows, feedlot operators demonstrated how the shared ideals of commercial feeders and agricultural researchers could be appropriated at public events to broaden the audience for their shared agendas.

It is also clear that Matsushima viewed his research as an extension of agro-industrial agendas as the two coevolved in the postwar era. During a broad-ranging interview in 2018, Matsushima—still mentally sharp at ninety-seven years old—provided effusive and detailed answers to questions about his research projects and interactions with commercial feeders. However, when asked to evaluate those relationships, Matsushima had little to say except, "We all worked together." When asked about whether there were ever any tensions between his research agenda and that of agro-industry, he revealed only that he received far more industry offers to conduct field trials and experiments than he could possibly complete and said, "I had to tell them what I thought."[47] What emerged from this conversation was a portrait of a skilled and tireless scientist who agreed that the primary value of his work was in advancing the goals of the commercial cattle feeding industry. Further, the timing of his career and his specific research projects reveal that the science and business of commercial cattle feeding coevolved, which explains in part why Matsushima never questioned industry goals or developed research projects independent of its needs. What is clear is that the growth and culture of commercial feeding on the Piedmont were bound up in Matsushima's csu career.

There is little doubt that efforts by Matsushima, Farr, and their feedlot and industry peers to transform commercial cattle feeding into a manufacturing operation prioritized the growth of the Piedmont's largest feeders over mixed farmers and small-scale commercial feeders. By the mid-1960s, agri-state marketing researchers argued that the feed mills employed by larger feeders were becoming essential tools of the trade. They also recommended power-boxes, which were self-mixing, self-unloading feeding units mounted on trucks—to save on human labor. These items were a significant expenditure, with the smallest feed mill costing over $11,000. Moreover, the rate of return for possessing such machinery was appreciably higher for feedlot operators who possessed more than one thousand animals.

Small feedlots of less than five hundred paid more per head for grain storage, feed pens, lighting, transportation, meatpacking, and land. They also struggled to obtain feeder cattle at prices comparable to their larger competitors and worked harder to convince processors of the quality of their product. Finally, their smaller size meant that they could only market fed cattle during limited periods of the year. This lack of flexibility made them vulnerable to price fluctuations. During the 1960s, when beef prices and demand steadily increased, small feedlots—possessing a capacity of five hundred head or less—usually succeeded. However, the principal growth in commercial feeding during those years came from feeders who maintained at least one thousand cattle at a time. Between 1953 and 1963, the number of cattle fattened on the Piedmont nearly doubled. At the same time, the percentage of cattle finished by small feeders dropped by 40 percent, while the quantity fattened by large feeders jumped by nearly the same percentage. During that same period, cattle feeding became the most valuable economic enterprise in the region, accounting for more than 40 percent of all Piedmont agricultural sales.[48]

Monfort and the New Piedmont Landscape

While agri-state scientists, agro-industry, hybrid corn, water projects, feedlot evangelism, and the Fed Beef Contest enable us to assemble the complex and multifaceted factors that transformed the Piedmont landscape, each of these interdependent parts was fully articulated in Monfort of Colorado. During the 1950s, Warren Monfort was at the forefront of operators who incorporated machines, chemicals, and pharmaceuticals into their operations. In addition, he both influenced and took advantage of the growth in corn hybrids, becoming the region's largest purchaser of grain and silage. In 1968 Monfort's feedlots could hold one hundred thousand cattle, making them the world's largest. He then built another feedlot south of Greeley, doubling his capacity. By the mid-1970s, Monfort of Colorado employed over two thousand people and fed six hundred thousand cattle annually.[49]

Monfort's operations reflected the scale and reach of a transformed Piedmont landscape. To meet the insatiable demand of Monfort's feed

Fig. 34. Cattle being vaccinated at Monfort Feedlots. City of Greeley Museums. Used by permission.

Fig. 35. Cattle being dipped at Monfort's Gilcrest feedlot. Once vaccinated, cattle swam through a chemical bath aimed at eliminating pests. City of Greeley Museums. Used by permission.

pens, the company's purchasers fanned out through the American West to buy an average of ten thousand steers weekly, transporting the animals on the company's fleet of orange-and-white trucks emblazoned with the "Monfort of Colorado" insignia. Their omnipresence and reputation for speed along the highway that connected San Francisco to New York prompted truckers to nickname the left lane of Interstate 80 "the Monfort Lane." At purchase, steer weights varied between six hundred and eight hundred pounds. However, the stress of confined highway transport dulled appetites and heightened susceptibility to disease. Consequently, feeders expected a small amount of weight loss in transit.[50]

Upon arrival at Monfort's feedlots, the full weight of animal science and mechanization were brought to bear to fatten each yearling steer to a slaughter weight of 1,100–1,250 pounds within 140 days. This was not a matter of merely placing them into pens and dumping an endless supply of corn into feed bunks. The trauma of being ripped from pasture grazing (the most common scenario), sold, and then transported dulled appetites for a ration they were not accustomed to eating. This was compounded by their exposure to BRD along the way. Consequently, after feedlot workers weighed, counted, and inspected the herd, they dispersed them into quarantine pens where they were numbered and socially distanced from their peers. Nutritionists and veterinarians checked steers twice daily for illnesses or abnormalities, adjusting feeds as necessary. After five to seven days of isolation, healthy animals underwent a feedlot "indoctrination program." This involved forcing animals through a specially designed concrete chute where veterinarians vaccinated them against BRD and other diseases and injected them with antibiotics that fought disease and accelerated weight gain. Needles extracted, cattle swam through a head-to-hoof bath of chemicals aimed at killing and warding off the various organisms that would otherwise thrive in the feedlot environment. Once run through the chemical and pharmaceutical gauntlet, feedlot workers herded animals into confined pens with other animals at the same stage. They had been moved three times since arrival. If all went according to plan, they would share a fenced, one-acre space with 450 of their peers for the next 140 days, until they had packed on four hundred pounds and were ready for slaughter.[51]

Obtaining slaughter weight depended on a mind-boggling quantity of feeds that were grown, harvested, transported, mixed, and delivered with jarring precision. Once confined, Monfort fed its cattle a diet that consisted of 60 percent roughage and 40 percent grains. While corn grain, a carbohydrate, facilitated the fastest weight gain, the work of industry and agri-state researchers determined that cattle's digestive systems required a transition period before they could be placed on "full feed." Locally grown corn silage was the roughage fed from October through May, with uncured alfalfa filling in the other four months.[52] After this transition period, steers spent the remainder of their lives consuming a feed ration consisting of 95 percent corn grain. The other 5 percent was a protein supplement mixed by Farr Better Feeds consisting of meat scraps, urea, soybean meal, several minerals, vitamin A, salt, DES, and antibiotics. Spread across Monfort's operation, these feeds amounted to 1.8 million pounds of roughage, 3.9 million pounds of corn grain, and 200,000 pounds of protein supplement each day. Monfort expected that steers would gain one pound of weight for every seven pounds fed, gaining between 2.5 and 3 pounds daily.[53]

Digestion of such a volume of feeds informed agricultural practice and altered the shape of the landscape. To supply its silage needs, Monfort contracted with nearly two hundred local growers within a twelve-mile radius each spring to plant over twelve thousand acres of corn to supply its seasonal need of three hundred thousand tons. Monfort also made one ton of cattle manure available to its growers for each ton of silage harvested. Eventually, manure volume outpaced demand and Monfort paid farmers to haul it away. Even with this boost, growers spent an average of twenty-five dollars per acre on synthetic fertilizers and chemicals to maintain their corn acreage and prop up a silage harvest of twenty tons per acre. While some growers harvested and delivered their silage, Monfort hired laborers to complete the majority of the harvest during a three-week period each September. For example, in 1974 Monfort employed twelve crews

Group of sight seers viewing the
filling of pit.

Fig. 36. Silage pit at Monfort's Gilcrest feedlot. The pit was 1,000 feet long by 200 feet wide, and 15 feet deep. At capacity, it held 150,000 tons of corn silage. City of Greeley Museums. Used by permission.

to harvest its silage contracts. In alternating twelve-hour shifts, they drove tractors and trucks into the fields to harvest, grind, and deliver much of the plant matter of the local region to Monfort's feedlots. Workers then stored the silage at Monfort's feedlots in massive pits capable of holding 150,000 tons, enough for eight months of feeding.

Gilcrest's pit silo was dug thirty-feet deep and occupied two hundred thousand square feet, or about the size of three and a half football fields. Influenced by Monfort, Weld County produced 42 percent of Colorado's corn silage in 1970.[54]

Monfort's sway over the growth of corn grain is more diffuse, but no less profound. To supply the company's 1.3-billion-pound annual demand, Monfort bought grain from farmers in Nebraska, Kansas, Iowa, and the Piedmont grown on 250,000 acres of land. The company maintained grain elevators in Kansas and Nebraska capable of holding 1.85 million bushels of corn and employed a steady stream of rail cars and trucks to transport the corn to the more modest elevators at its feedlots. While the company prioritized Midwestern grains over those cultivated on the Piedmont, Monfort's smaller feedlot peers, such as W. D. Farr, utilized local corn. Moreover, between 1964 and 1974, the production of corn grain on the Piedmont increased by 150 percent, and nearly every kernel was consumed by cattle.[55]

Whether transported from five hundred miles away or just up the road, corn required one more transformation before its journey to the feed bunk. As with so much of the feeding industry, collaboration between agri-state and industry was critical. During the late 1950s, as John Matsushima was eating breakfast with a cattle feeder, they discussed how to best take advantage of feed gains from corn grains. As Matsushima tells the story, he was struck by the number of people eating corn flakes and how the process of flaking corn made the carbohydrate energy from the grain more readily available to humans. Thinking that this process might support more efficient digestion in cattle, Matsushima got to work on a process for flaking and then steaming feedlot corn, perfecting it in 1964. Two years later, with Matsushima's guidance, Warren Monfort integrated steamed corn flakes into cattle diets. The process employed rollers that flaked the corn and boilers that cooked the grains for eighteen minutes at two hundred degrees Fahrenheit. Computerized scales then weighed the steaming hot grains and divvied them up according to the demands of each feedlot pen, finally dumping them into a waiting truck where they were mixed with silage and a protein supplement. Monfort trucks mixed feeds en route to each pen, depositing the rations still hot from the boilers. In this fashion, Monfort fed nine hundred tons of steam-flaked grain per day. According to Matsushima, hot corn flakes increased feeding efficiency by 10 percent. Warren Monfort's son, Kenny, argued that steam-flaked corn was the greatest advance in the industry since DES. Even Kellogg's, America's largest manufacturer of a corn flakes breakfast cereal, must have been impressed, since Monfort's cattle alone ate four times more corn flakes daily than did Kellogg's human customers.[56]

After a feeding period averaging 140 days, living steers unwittingly made their final journey—to the slaughterhouse. Prior to 1960, that journey involved being packed into cattle truck trailers and train-cars "on the hoof" and freighted to one of the nation's large terminal markets in Chicago, Omaha, Kansas City, or Denver, where steers would be killed, processed, and shipped to consumers. However, in 1960, as Monfort's Greeley feedlot approached twenty-five thousand head capacity, Kenny Monfort convinced his father that building an in-house packing facility could be profitable while enabling the company to ride out downturns in a volatile beef cattle market. At the time, nearby Denver's meatpacking industry was growing. Consequently, in 1960, Kenny partnered with Denver-based Capitol Pack to build a slaughterhouse adjacent to Monfort's Greeley feedlot. A year later, when Kenny Monfort sought to expand packinghouse operations, Capitol balked at the expense. So, Monfort bought out Capitol's share. The new ninety-two thousand square foot packing-house, renamed Greeley Pack, was the largest facility of any kind constructed in the Greeley area since Great Western built its sugar factory in 1902.[57]

Initially, Greeley Pack removed hides, head, and offal and cut ani-mals into sides of beef, then shipped those sides to supermarkets, restaurants, and other clients, where butchers processed beef further by removing additional fat and portioning the sides into recogniz-able cuts. When cattle reached the target weight of approximately 1,150 pounds, Monfort employees on horseback drove the steers to a packing pen, ninety minutes before slaughter, then herded them into

Fig. 37. Kenny Monfort (*center, dark jacket*) inside the Greeley Pack Slaughter-house. City of Greeley Museums. Used by permission.

the "knock chute" where a retractable bolt gun killed each animal. Employees then shackled each steer by its left hind leg to be lifted and hung on a moving assembly line. The next employee slit the throat, initiating the bleeding process. Once the bloodletting was complete, the animal moved from the blood pit to the skinning area, where several employees painstakingly removed animal hides. After skinning, workers sliced through the still-warm flesh, removing the offal. Workers then sawed the carcass into two sides, covered them individually in cheese cloth that had been soaked in a light brine solution, and moved them into massive coolers. The death-to-chill assembly line occupied twenty-six minutes. Twenty-four hours later, Monfort workers removed the cheese cloth and broke sides between the twelfth and thirteenth rib, so that USDA employees could visually examine and grade carcasses. Monfort boasted that 85 percent of its steers received the desirable USDA Choice grade.[58]

While preferable to shipping on the hoof, transporting chilled sides of beef still resulted in wasted space and unrealized profit. Further processing could fill industrial coolers and trucks more efficiently. Moreover, processed cuts of meat netted more profit than selling sides of beef. Consequently, in 1965, Kenny Monfort became the first commercial cattle feeder to integrate beef fabrication into packinghouse operations. Principally, fabrication involved removing additional fat, breaking beef sides into primal and subprimal cuts, deboning the majority of the carcass, and processing less-valuable cuts into ground beef. Monfort was particularly innovative in the arena of boxed beef. Once butchers completed their disassembly of each carcass, Monfort's automated system vacuum sealed individual cuts of meat in plastic, boxed them for delivery, and added dry ice pellets to each container. Once sealed, Monfort used automated loaders to fill waiting trucks, leaving virtually no wasted space. The packaging process also had the advantages of extending shelf life and maintaining the red meat coloring consumers in self-serve grocery markets used (falsely) to determine the freshness of their purchases. By the early 1970s, 90 percent of all cattle in Monfort's feedlots exited its packinghouse in boxes. Describing the industrial processes at the packing plant, Greeley feedlot manager Duane Flack stated that the packing plant was "an endless maze of conveyors, and sophisticated electronic handling systems with elaborate communications devices [that made] Monfort's packing plant look more like one of Detroit's automotive assembly plants than a meat packing facility." By design, assembling cars and disassembling steers mimicked each other.[59]

Boxed beef fully articulated W. D. Farr's admonishment that manufacturing beef was the economic and cultural paradigm through which feeders should view their operations. The boxes filling a loaded Monfort truck were indistinguishable from those exiting most factories in America. In form and appearance, the beef inside each box bore no resemblance to the biological processes that made it. Boxed beef also represented the fulfillment of John Matsushima's aspirations in the Fed Beef Contest. The outward appearance of living animals and pride of ownership were inconsequential. A steer's value became apparent only after slaughter, when steers received a USDA grade. Since most boxes on a Monfort truck contained beef that graded Choice, the company embodied a scaled up, vertically integrated version of Matsushima's philosophy. The culture of packing uniform boxes full of beef aimed at the consumer palate transformed the thinking of Piedmont commercial feeders who sold fattened cattle to Monfort or one of its packinghouse competitors. In describing the feeding that took place on his family's Piedmont operations, Richard Seaworth recalls that he timed the sale of his cattle to the packinghouse with a view to making sure each steak "fit inside the box." Seaworth argued that consumers wanted eight-ounce Choice steaks, and that his selection of breed, feeds, and time of slaughter would yield "one extra steak [in each box] that grades the way I want it to."[60] According to Monfort, every steer carcass yielded seven boxes total of chuck, rib, round, and loin steaks. Seaworth, like Monfort, Farr, Matsushima, and their peers, learned to view individual steers as boxes of beef and feedlots as full truckloads.

The contents of every box and the destination of each truck linked Monfort's beef factories together with larger agro-industrial networks. Just as the carbon byproducts of the oil and gas industry were employed to synthesize the petrochemicals used to kill weeds and pests that attacked crops and cattle, so they were also critical to transporting beef. Monfort hired the Dow Corporation to design the petroleum-based plastic bags used to vacuum seal its various cuts of meat. For specialty cuts and nonstandard orders, Dow designed Saran Wrap S Bags of various sizes that formed around the meat prior to being "automatically conveyed through a shrink tunnel." For the minority of customers who ordered whole sides of beef or hind-quarters, Dow supplied plastic lining for Monfort's boxes. Fossil fuels also kept the meat cold in transport. The company, Liquid Carbonic, supplied the dry-ice pellets that filled out every Monfort box. As the solid form of carbon dioxide, dry ice is a petroleum byproduct. In this case, Liquid Carbonic synthesized the refrigerant from the left-overs of its parent company, Houston Natural Gas.[61] Boxes, dry ice, and vacuum sealing opened broader distribution doors since they

Fig. 38. Machine processing beef inside the Greeley Pack Slaughterhouse. Once portioned and trimmed, individual steaks were placed on an automated conveyor belt (seen here), where each cut was vacuum sealed, packed in dry ice, and boxed before being packed on a truck for transport. City of Greeley Museums. Used by permission.

Fig. 39. Boxes of beef awaiting transport inside the Greeley Pack Slaughterhouse. City of Greeley Museums. Used by permission.

enabled Monfort to transport and chill more beef. Consequently, in 1968, it merged its distribution operations with Denver-based Mapelli Brothers Food Distribution Company. The partnership enabled Monfort to build twenty-four food distribution locations in the United States and sell its product overseas. By the early 1970s, Monfort was Japan's largest foreign beef supplier.[62] In the forty years since Warren Monfort started feeding cattle year-round, Monfort of Colorado had become a multinational corporation that voraciously consumed the resources of the Piedmont and beyond, distributing its manufactured output far and wide.

Unbeknownst to consumers buying the typical Monfort steak—unloaded from a truck, pulled from a box, and placed in the refrigerated section of the market—they were eating more than beef. That inexpensive steak was a product of engineering and biological and chemical manipulation. The corn grain and silage that drove cattle rations were only available in quantity as a result of corn hybrids that increased yields everywhere and made cultivation viable on the Piedmont. Corn economics, however, required more than just good seeds. Massive infusions of synthetic nitrogen and herbicides enabled farmers to supplant soil ecology with chemical technology. During the 1960s, this was increasingly true, due to a new organic biocide named atrazine that effectively eliminated the broadleaf weeds that choked out corn crops. According to Piedmont farmers who supplied grain and silage for cattle feeders, the chemical not only eradicated weeds, but its persistence rendered a field useless for crops other than corn for several years, making corn both attractive and addicting.[63] An atrazine fix made monocropping a chemical and biological imperative. The cheap Monfort steak was also a pharmaceutical byproduct. Antibiotics limited diseases endemic to feedlots, moved animals more quickly onto feeds they had not evolved to eat, and boosted efficiency. Meanwhile, hormones such as DES accelerated growth rates. Feedlot and packinghouse engineering added a final layer to cheap beef. Monfort's finely tuned operations eliminated inefficiencies by tightly regulating the final months of a steer's short life while managing the supply of feeds entering its lots and the meat that exited. The typical Monfort steak that emerged from a box was a marvel of biological, chemical, and technical innovation that reengineered the Piedmont landscape.

Piedmont commercial feeders and their support system of agri-state scientists and industrial allies shattered long held notions of value and temporality. Their research, field trials, organizations, and public events were all aimed at what W. D. Farr called "manufacturing beef."[64] As the efforts of Farr, Monfort, Matsushima, and their peers suggested, cattle should not be viewed as products of biology or domesticated animals with deep attachments to humans. Rather, they existed as inputs for an industrial system aimed at manufacturing a desirable commodity. Assembly line beef also compressed and bypassed nature's processes. During the early twentieth century, the lifecycle of the typical steer or heifer, bred for meat, was three to four years. Commercial feeding cut beef cattle longevity in half. Further, their truncated birth-to-box lives existed in inverse proportion to expanded corn grain and silage yields. The critical quantities of corn required could not be cultivated with existing practices of crop rotation and mixed farming. The bevy of synthetic biocides and fertilizers tested and manufactured after World War II enabled farmers to replace the slow-release restoration processes supported by traditional practices with the on-demand results propelled by agro-industry. By reengineering nature's processes or ignoring them altogether, commercial cattle feeders reconfigured biological clocks and cultural values.

Monfort of Colorado represented the triumph of the new Piedmont agroecology. From the settlement of the Union Colony in 1870 through World War II, soil health was the keystone of Piedmont agriculture, and Piedmont farmers harnessed the region's plant matter and associated organisms to support it. Nutrient cycling laid the beet sugar industry's foundation, supplied the principal rationale for maintaining cattle, and was the cornerstone of long-term financial stability. Paired with irrigation, biologically sustainable soils supplied the agricultural energies of an entire region. Commercial cattle feeding supplanted soil as the object of Piedmont agroecology. In the feedlot, soil became dirt—mere matter, compacted to hold bovine weight. Moreover, feedlot operators directed all of their operational trappings—pens, mills, feed bunks, and drinking troughs—at the rapid physical growth of cattle. For crop farmers supplying cattle feeds, soil existed primarily to hold corn plants in place as organic biocides and chemical fertilizers supplied nutrients and killed terrestrial and airborne pests. Cattle became both factories and food, enabling the Piedmont to transform postwar chemical, biological, and pharmaceutical innovations into precision cuts of beef calibrated to satisfy consumer demand.[65]

Yet, this gives Monfort of Colorado and its commercial feeding peers too much credit. In many ways, the rapid rise of concentrated cattle feeding on the Piedmont was the logical extension of an economic philosophy present for a hundred years but held in check by an agroecology of the possible. Efforts by Piedmont cattle feeders, industry-energized innovations in the chemical and biological sciences, and the narrow and unquestioned support of the agri-state research complex released those reins. They curbed or eliminated entirely the need for crop rotation and mixed farming, transformed fields into feedlots and organisms into pests slated for eradication, and created production timelines that synchronized with market demands. In that sense, Monfort represented not so much a radical departure but the full articulation of an unquestioned industrial ideal that had existed for a century.

Perspective

Kent Peppler is a fourth-generation Weld County farmer with a story that resonates with Piedmont history. His father's side of the family was German Russian immigrants who immigrated to Colorado in 1907 and followed the typical progression from sugar beet laborers to farm tenants to owners, eventually settling in 1945 on land near the town of Mead. Peppler now farms this land. His mother's side of the family, imbued with frontier dreams, relocated from Detroit to Mead in 1910, bought land and cultivated a diverse set of crops that included sugar beets contracted to Great Western Sugar. Peppler's grandfather fed sheep through the 1930s, adding cattle in 1939. Growing up during the 1960s and 1970s, Peppler recalls feeding livestock with beet byproducts and a portion of the barley crop that his family sold to Coors Brewery. He remembers "force feeding antibiotics" and diethylstilbestrol (DES) to fight off disease and promote rapid growth. When Peppler began to farm on his own in 1981, he continued to rotate crops such as corn, sugar beets, alfalfa, and barley, while supplying some of his land's nitrogen and phosphorous needs from cattle manure. However, as Peppler explains, he quickly recognized that his cattle-feeding days were numbered. A mixed farming operation such as his, with only five hundred head of cattle, simply could not compete with the economies of scale and industrial efficiencies exemplified by neighbors such as the Monforts. In fact, Peppler recalls a 1983 conversation with Kenny Monfort in which the cattle baron predicted that, by the year 2000, those not feeding at least fifty thousand animals would soon be out of business. In debt, Peppler abandoned cattle in 1990. Pests and cultivation costs forced him out of sugar beets soon after.[1]

When I met Kent Peppler for a tour of his Piedmont farm in June 2019, an ominous sky crackled with the sound of summer thunderstorms. Peppler is stout, with brown hair and youthful features that belie his sixty-plus years of life. He possesses an easygoing, aw-shucks mannerism that invites conversation. The majority of Peppler's lands are divided between corn and barley. Like his father, he sells barley to Coors and to local dairies, while marketing corn grain and silage to agricultural cooperatives who contract with commercial feeders. Though Peppler practices some crop rotation, most of his corn is grown on the same land annually, using genetically modified Roundup Ready seeds and herbicides. Though Peppler has some misgivings about the environmental problems associated with mono-cropping, he states that the combination of efficient weed removal and consistent crop production associated with genetically modified corn inform his cultivation decisions. Moreover, Peppler can use a single tractor and several attachments to manage hundreds of acres without hiring additional labor.[2]

Despite these efficiencies, Kent Peppler explained that crops do not provide sufficient income. So, he leases farmland for another purpose. As we traveled from one field to another, a series of sand-colored cylinders stood amid otherwise perfect rows of corn. Several stories tall and visible from half a mile away, these tank batteries were constructed by Crestone Peak and Anadarko, oil and gas companies, to store and process the abundant reserves of shale oil that exist thousands of feet below Peppler's lands. As Peppler explained, he leases some of his farmland to these companies who use it to drill for and store oil and natural gas. In exchange, Peppler receives revenue based on the value of the unearthed fossil fuels. Each well is unlocked through a process called fracking, which uses petrochemicals, sand, and between 1.5 and 16 million gallons of water at high pressure to fracture bedrock thousands of feet below the earth's surface, freeing existing reservoirs of oil and natural gas. Peppler could raise between two and twenty-five acres of corn with the water used to frack a single well.[3] Oil and water may not mix, but they are mutually dependent in this case. So it was that on this day, just as the impending thun-

Fig. 40. Hydraulic fracking wells in Kent Peppler's fields. Photo by author.

derstorm began to unleash its fury, Peppler stopped to chat with an oil worker, hoping for news of an abundant oil harvest. Peppler later explained that most small- to mid-sized farmers like him depend on the oil and gas industry for their ongoing economic health. It is one of the factors holding Piedmont agriculture in place.[4]

While the presence of the fossil fuel industry on Kent Peppler's farm presents as a divergence from his family's agricultural past, there is an abundance of historical continuity. Beginning in the 1920s and accelerating after World War II, the oil and gas industry was hidden in plain sight on the Piedmont. The organic chemicals used to disinfect seeds and kill weeds and insects that targeted crops and livestock were synthesized with carbon molecules that were byproducts of the oil and gas industry. Previously stored in the earth's crust for millions of years, these same fossil fuels also powered the growing numbers of tractors and attachments that made cultivation and feeding more efficient. Despite oil's ubiquitous presence, prior to the advent of fracking on the Piedmont in the twenty-first century, those fossil fuels were extracted, refined, and transported from elsewhere. As Kent Peppler's experience illustrates, they are ever present on the Piedmont, but not just as energy to cultivate crops. For farmers such as Peppler, fracking the petroleum locked thousands of feet below the earth makes agriculture on its surface possible.

Peppler's decision to support the cultivation of market crops using every available technology and resource resembles decisions made by Union colonists. Nineteenth-century Greeleyites were committed to what historians have referred to as intensive cultivation. Rooted in a market-based philosophy, intensive cultivation refers to "the application of capital and technology to increase yields on existing land," as opposed to extensive cultivation that expands production by farming additional lands with existing technology.[5] Greeley settlers believed that engineering the region's waterways to irrigate their fields would generate abundant harvests of marketable crops, and modest amounts of wealth. Along with local boosters and other farmers in the area, they courted rail energies to access agricultural markets. In the late nineteenth century, Piedmont farmers integrated biological and chemical sciences through their deployment of nitrogen-fixing alfalfa and by raising quantities of livestock calibrated to produce needed manure for fertilizer. It was a relatively productive and some-what sustainable agriculture that employed the resources and existing technology at their disposal.

The industrial agriculture imported onto the Piedmont by Great Western Sugar did not replace intensive cultivation; rather, it was an extension of it. Using that logic, it makes perfect sense that most Piedmont farmers embraced the economies of scale that resulted from the beet sugar industry in the early twentieth century. After all, sugar beet agriculture required meticulous soil maintenance and offered opportunities to make more money on existing acre-age. Though some farmers flinched at the idea of an imported labor force, seasonal labor was the devil's bargain they made to advance the farming they embraced. Intensive cultivation maps onto Piedmont irrigation just as well. Throughout the period under study, farmers sought more water, recognizing that water meant wealth, predictable yields, and more lucrative crop choices. The year-round feedlots that materialized in the 1930s took advantage of new feed technologies, organic chemicals, and the hybridization of corn that enabled it to be grown cheaply and abundantly on Piedmont farms. When farmers adopted petrochemicals as a centerpiece of farm management, they abandoned former methods, but not prevailing philosophies. At the heart of chemical agriculture was the belief that modest capital investment in technology could minimize human labor and facilitate vastly increased yields on existing acreage. Union colonists did not have access to the kind of capital that these modern farmers did, nor would they have recognized their machines and chemicals. But they would have identified the intensive cultivation philosophy behind it.

There is also continuity in the philosophies that guided agri-state scientists and officials from the late nineteenth century through the 1960s. Their presence on the Piedmont was supposed to be guided by two missions: to support independent farmers and to conduct and promote scientific methods. There is little evidence that the former mission ever drove their work. Instead, when agri-state researchers thought about independent farmers, they blithely assumed that new technologies universally benefited all farmers and further, that expanded yields on existing acreage were an unquestionable good. During the late nineteenth and early twentieth centuries, prioritizing technology over independent farming was of little practical consequence since the research necessary to expand yields involved soil maintenance, irrigation engineering, and integrating cattle feeding with crop rotations. Most CAC and USDA research during that period occurred either on experiment station lands or on those of farmers themselves. However, when technology shifted toward machinery and chemicals, beginning in the 1920s, agri-state science shifted with it—away from Piedmont farmers and soils and toward agro-industry. This was not so much a modification of philosophy as a change in relationships. As agro-industry promised greater efficiency and higher yields through broad application of new technologies, agri-state researchers gravitated toward industry-generated research, whether in fields or commercial feedlots. This should not come as any great surprise since supporting technology in agriculture had always been central to agri-state's mission.

It is possible to imagine other outcomes. Prior to the 1920s, the vast majority of agricultural research on the Piedmont integrated well with the form of farming that already existed, focusing on soil maintenance, crop rotation, and mixed farming. Regardless of philosophy, the same research that encouraged higher yields, also supported sustainable farming. When the products of agro-industry became readily available on the Piedmont, the complex of agri-state researchers and officials failed to question the basic assumptions inherent in their philosophy. New technologies did not work with existing forms of agriculture; they supplanted them, suggesting that capital investments in machines and chemicals would make traditional farming irrelevant. The new products boldly stated that it was not necessary to use alfalfa and manure to replace essential nutrients and all hand labor could be replaced by precision machinery and biocides. Farmers did not need to wait for an insect outbreak before using insecticides, since the new products claimed that they could kill the offending organisms before they invaded fields. Moreover, lethal chemicals precluded farmers from worrying about the unwanted pest regimes that could result from monocropping and year-round feedlots. On the Piedmont, the new agriculture suggested that soil could simply be a platform for cattle to stand on or, as historian Colin Duncan has aptly put it, "Merely a physical support system for [plant] roots." The new agriculture did not assert that soil could be more productive through technology, but that the soil—as an interconnected biological unit—could be made irrelevant through technology.

These revolutionary transformations in agriculture demanded a radical reorientation in agri-state philosophy and research agendas. Until the 1960s, when a critical mass of ecologists began to question the trajectory of U.S. agriculture, the USDA and agricultural colleges possessed the only complex of publicly minded researchers within agriculture in a position to question the long-term consequences of the new agriculture.[6] However, one of the most glaring features in agri-state research was its failure to conduct basic research related to modern industrial agriculture. Virtually no work was done by scientists to examine how chemicals impacted relationships among organisms in the soil or the influence of machinery and chemicals on local watersheds. Researchers were concerned about the fact that chemicals killed insects but conducted little research into insect toxicology or why certain populations developed resistance. The agricultural transformations that occurred through the middle third of the twentieth century failed to spark a critical self-evaluation within state-sponsored science.[7]

Even for those within the academy who questioned its direction, few pathways existed. Consider the career of Robert Zimdahl, emeritus professor of weed science at CSU. Zimdahl was hired in 1968 after a graduate and professional career that included operating a chemical spraying business in upstate New York, performing herbicide research for Pepsi, and completing advanced degrees at Cornell and Oregon State. Despite lucrative offers from Monsanto, Shell, Diamond Alkali, and Geigy, Zimdahl chose a faculty position at CSU, in part because of its location near the Rocky Mountains. During his initial years there, Zimdahl received funding from Great Western Sugar as well as $5,000 from Coors to complete a study of how herbicides could kill off the wild oats that threatened sugar beet plants and the barley used by Coors to brew its beer. Describing the nature of his research, Zimdahl explained, "It was interesting work, but it was not based on some fundamental hypothesis. . . . Back then it was a lot of testing of the big herbicides [such as] 2,4-D, atrazine, and dicamba."[8]

If Zimdahl's critique of conducting field trials for industry was correct—they lacked fundamental scientific hypotheses. Then why would a professional scientist choose to devote so much time to them? According to Zimdahl, money played the critical role. While plenty of money flowed into university coffers through formal arrangements between industry and the institution, more money found its way into the hands of CSU researchers when agro-industry contacted scientists directly. According to Zimdahl, representatives of a chemical company would "show up at your door and say, 'We have this product

Fig. 41. Photograph of Robert Zimdahl. University Historic Photograph Collection, CSU Morgan Library Archives and Special Collections. Used by permission.

and we would like to have you test it . . . go out and spray this stuff on these crops at this rate at this time and tells us what happens. Then tell us what it does to this crop and what it does to the weeds' . . . and they'd give you money." Zimdahl went on to describe how two of his colleagues, plant pathologist Jess Fults and weed scientist Gene Heikes, cultivated relationships with petrochemical companies that underwrote their careers. That money typically came with a significant carrot. Industry was generally not interested in an itemized list of how their funds were spent, as long as the research was conducted professionally and completely. Consequently, researchers commonly applied excess funds to their own research. If a researcher wanted to hire a graduate assistant and the funds were not forthcoming, then industry research could pay the tab. Or, as Zimdahl pointed out, "You needed some dependence [on industry] to gain some independence."[9]

The same agrochemical funds that presented research carrots were also sticks. Zimdahl explained that it was impossible for a weed scientist or an entomologist to succeed professionally aside from cultivating relationships with industry. In weed science, no funding existed for any kind of research other than that which employed chemical herbicides. According to Zimdahl, had he chosen any sort of "alternative agriculture . . . , [he] would have failed because there was not any money for that kind of agriculture." So, a sort of path dependence developed whereby the predominance of biocides in agriculture since World War II and the money that the chemical companies brought to the table determined the available options for those in agricultural research.[10]

According to Zimdahl, ethics in his field revolved solely around "what we [could] do, not what we ought to do." In other words, weed scientists and entomologists desired only to know whether a biocide did what it claimed to do. Further, agri-state scientists believed that the most significant consequence of their research was the production of more food, and that more food was an unqualified good. It followed that those who questioned the chemical paradigm obstructed food production. Regulators in the FDA and ecologically minded scientists, such as Barry Commoner and Rachel Carson, who were concerned with chemical harm, hindered the agri-state mission. According to Zimdahl, his colleagues felt that making a moral dilemma out of petrochemicals "[did] not lead anywhere helpful."[11]

One of Robert Zimdahl's transformative professional experiences illustrates this narrow thinking. In the early 1970s, Zimdahl began to question the pervasiveness of chemicals in agriculture. He presented a paper at a conference for the Weed Science Society of America in Dallas titled "2,4,5-T—A Value Question." Though 2,4,5-T was employed in clearing unwanted vegetation from fence lines and public rights-of-way, it was most well-known in the early 1970s as one half of the compound Agent Orange, used to defoliate dense forests during the Vietnam War.[12] Though Zimdahl did not advocate for a ban on 2,4,5-T, he expressed concern over its toxicity. When Zimdahl finished speaking and left the room, he was confronted by six weed scientists. After giving him the obligatory congratulations, they pointed to what they felt was the central problem in his presentation. Zimdahl lucidly recalled them stating, "And the problem is you." They went on to argue that by questioning a proven herbicide, Zimdahl was casting doubt on the tools of their profession and, by extension, siding with the "environmentalists" who wanted to "take [their] tools away." Later, when he submitted his findings for publication, every weed science journal rejected them. Zimdahl finally published his conclusions in the *Bulletin of the Entomological Society of America*.[13] By posing a moral dilemma to scientists solely interested in chemical efficacy, Zimdahl began to operate outside of the knowledge economy his profession established for itself.[14]

Though, as Zimdahl points out, state-sponsored science failed in its role as guardian of independent farmers, we cannot disentangle that failure from a rapidly industrializing agriculture in the twentieth century. At the heart of industrial production, whether in a factory or on the farm, is the desire to rationalize nature for economic gain. This is arguably more complex on farms than in factories. Brick-and-mortar factories manufacture goods from an amalgamation of minerals, fossil fuels, chemicals, and machinery. Production sites are generally located in urban areas where a critical mass of laborers is available to work for wages. In the farming that predominated on the Piedmont prior to World War II, production depended on biological and climatological factors that were largely absent from physical factories. Healthy soils resulted from a complex interdependency of organisms and minerals. Farmers had to maintain enough crop diversity and use on-the-farm resources to replenish nutrients that heavy cropping removed. There were always limits to what could be grown, and farmers ignored the complexities of their land at their own peril. But despite the most meticulous management, unusual weather, pest infestations, or lack of available water could curtail production in any given year. Complexity and a degree of unpredictability dulled accurate predictions. Labor also curtailed efforts to rationalize agriculture. It was never easy, even for corporations like Great Western, to attract a reliable and consistent seasonal labor force willing to do hand labor solely during periods when crops required it. To supply a demanding market with human food and animal feeds, farmers and industry developed complex methods to simplify or supersede biological relationships. In that sense, irrigation projects, petrochemicals, pharmaceuticals, and machinery were vastly different tools with a common goal: to rationalize nature's unpredictability. That required simplifying the agricultural landscape, and that could not be done without doing violence to biological relationships.

Simplified landscapes have also altered the human landscapes of the Piedmont. Since World War II, chemicals, machinery, and water rights have been the most instrumental factors in farming success. All of these require heavy capital investments. Postwar hybrid corn cropping presents an instructive example. Corn required more water and biocides than any staple crop that preceded it. Moreover, market farmers could not grow it without tractors and fossil fuels. Farmers without sufficient water rights—even with the new water from the Colorado-Big Thompson Project—or the ability or inclination to invest in machines and chemicals could not grow the most lucrative crop. Statistics from 1933 to 1959 show that in Weld County, the most productive county on the Piedmont, farm size more than doubled as wealthier farmers bought out those who could not afford to stay in business.[15] That trend continued in the commercial feeding industry in the years that followed. The number of feedlots in Colorado peaked at nearly 1,400 in the late 1960s. By 1991 that number had dropped below three hundred. Kent Peppler was among those cattle casualties. Yet, the number of feedlot cattle increased. Bill Hamerich, president of the Colorado Livestock Association (formerly CCFA), states that in 1982 his organization had 480 member feeders. Today, the number of cattle fed by his members remains roughly the same, but membership has declined to eighty-two.[16] In her seminal work on industrial agriculture, Deborah Fitzgerald explains that new agricultural technology always began as something that a few wealthy farmers purchased to make their operations more efficient, but it quickly became a tool of necessity that forced some farmers to go deeply into debt or get out of farming.[17]

There is no easy answer to the quandaries posed by industrial agriculture in the modern world. While my sympathies are with agricultural philosophers such as Wendell Berry, who believes that it is possible to return to a preindustrial agriculture whose principal energies are supplied by draft animals, my idealism is tempered by present realities.[18] If the history of Piedmont agriculture offers any wisdom for the present, it is that food production must be characterized by restraint, patience, and esteem for the values of local peoples and local production and consumption. Land ownership, farmer income, and food production on the Piedmont was at its most stable from 1900 to 1930, when small acreage, mixed farming, and minimal use of technology were employed. The largest tragedy of that era was the importation and marginalization of seasonal laborers. However, if reformers such as Thomas Mahony were correct, the profit margins achieved by Great Western Sugar and its growers could have supported living wages for hand laborers without altering farm production. That same period was also characterized by the use of local feeds and by the recycling of most crop wastes back into the soil. While all farming disrupts ecological relationships in the very act of inserting a plow into the soil, farmers on the Piedmont were limited largely to a form of farming that relied on the resources of their own farm, or those within a small radius of their property. The greatest damage to human values and ecological relationships occurred during the immediate post–World War II era when farmers and agri-state scientists adopted industrial products without questioning their long-term harms, homogenizing what had previously been a diverse agriculture and embracing unfamiliar and extra-local inputs.

It is that apparent failure to interrogate potential harms and consequences that resonates most, since it lacks an ethical core that extends beyond economic efficiency. After all, how a society feeds itself is fundamental to its health, in the broadest sense of the word. Consequently, its agriculture should reflect broad and inclusive ethics whose burdens are shared by producers and eaters alike. Healthy agriculture—industrial or otherwise—should be marked by respect for the complexity of biological relationships in the land; attention to how crops, animals, and their by-products can be consumed and reused locally; restraint in the use of technology; and a larger society that respects the foregoing goals. Historian Donald Worster observed that good farming should make people healthier, promote justice, and preserve the earth and its network of life.[19] These are goals still worth striving for.

NOTES

INTRODUCTION

1. "Portrait of a Packer-Feeder"; Roe, "They Feed 200,000 Steers a Day"; "Packing Plant Slide Show," unpublished typescript, JBS box 6, Scripts.
2. David Boyd, *A History*, 39–74.
3. See for example Pollan, *Omnivore's Dilemma*; Kingsolver, Hopp, and Kingsolver, *Animal, Vegetable, Miracle*; Berry, *Unsettling of America*.
4. For an excellent example of nature pushing back against human efforts to rationalize agricultural production during the late nineteenth and early twentieth centuries, see Fiege, *Irrigated Eden*.
5. Wyckoff refers to the Arkansas River Valley as the Southern Colorado Piedmont. For the sake of narrative quality, I will use the term "Piedmont" to refer to the Northern Colorado Piedmont. For a history of the Southern Piedmont in the twentieth century centered around the Dust Bowl, see Sheflin, *Legacies of Dust* . For Wyckoff's description of the Northern and Southern Colorado Piedmont, see Wyckoff, *Creating Colorado*, 101–53.
6. Wyckoff, *Creating Colorado*, 101–53. By contrast, the Columbia River, the largest river in the American West—admittedly, draining a much larger landscape, delivers 190 million acre-feet annually. An acre-foot is the amount of water necessary to cover one acre of land in one foot of water. It is the measurement most commonly used by farmers to describe their water appropriations. Consequently, the acre-foot will be the predominant unit of measurement employed.
7. Wyckoff, *Creating Colorado*, 101–53; Jessen, *Railroads of Northern Colorado*.
8. Books that examine water in the American West post–Civil War include Pisani, *To Reclaim a Divided West*, and Worster, *Rivers of Empire*.
9. Those who examine Piedmont water development in the initial decade after Greeley's settlement include legal historians David Schorr and water law expert Greg Hobbs. Both argued that Greeley's approach to irrigation was motivated by finding practical and legal remedies to the problem of aridity. By contrast, environmental historian Donald Worster argues that Greeley

settlers' approach to owning water rights was motivated less by the need for water and more by a speculative impulse to exploit nature for profit. In this, there is continuity with historians such as Ted Steinberg who argues that textile manufacturers in New England had been monopolizing the public waters of New England for private gain since the early nineteenth century. See Schorr, *Colorado Doctrine*; Hobbs, *Public's Water Resource*; Worster, *Rivers of Empire*, 83–96; Steinberg, *Nature Incorporated*.

10. On Colorado's use of irrigation in relationship to the rest of the West in the early twentieth century, see Hemphill and United States, *Irrigation in Northern Colorado*; Laflin, *Irrigation, Settlement, and Change*.

11. White, *Organic Machine*.

12. Studies of agriculture in California include Stoll, *Fruits of Natural Advantage*; Walker, *Conquest of Bread*; Pisani, *From the Family Farm to Agribusiness*; Worster, *Rivers of Empire*.

13. Even at the end of the Civil War, when the Piedmont possessed no agricultural settlements of note and the Homestead Act was still an unrealized dream of the Republican Party, land costs in California were already running as high as ten dollars per acre. See Cochrane, *Development of American Agriculture*, 78–98.

14. Cited in Walker, *Conquest of Bread*, 3.

15. Crops that were adopted in California, receded in the face of more lucrative cultivars including wheat and alfalfa. These remained staples on the Piedmont and throughout the West throughout the first half of the twentieth century. For a general description of the adoption of wheat in the West, see Cochrane, *Development of American Agriculture*, 78–98. For historical case studies of wheat in the West, see Fitzgerald, *Every Farm a Factory*, 129–56; and Duffin, *Plowed Under*. On alfalfa in the Snake River Valley of Idaho, see Fiege, *Irrigated Eden*, 143–55. On alfalfa in Colorado, see Steinel and Colorado Agricultural College, *History of Agriculture*, 413–18. Donald Pisani provides an example of the persistence of alfalfa in California in the late nineteenth century to feed livestock on the mega-ranch of Charles Lux and Henry Miller. See Pisani, *From the Family Farm to Agribusiness*, 191–249.

16. In 1944 Americans consumed 55 pounds of beef per capita. In 1969 they consumed 110 pounds. See W. D. Farr, "Cattle Feeding in the United States— Its Present and Future" (speech, San Antonio TX, December 4, 1969), W. D. Farr Collection, box 1.

17. For an excellent example of a book that approaches agroecology from the perspective of global commodities, see Soluri, *Banana Cultures*.

18. Worster has argued that capitalist regimes encourage the exploitation of nature so that individuals can accumulate wealth. He further argues that,

prior to 1930, the very idea that the health of the land should constrain human activity was abhorrent in the United States. In his analysis of commercial chicken feeding operations, William Boyd argues that nature was viewed by agro-industry as a series of constraints to be overcome through technological and scientific expertise. In his analysis of Great Plains farming in the twentieth century, Geoff Cunfer emphasizes the region's agroecological volatility even as farmers were able to sustain the productivity of their farms over long periods of time. Others, such as Bart Elmore, in his analysis of Coca Cola's business practices, have argued that successful capitalist regimes indirectly exploit nature by demanding cheap raw materials and thus forcing suppliers to jettison environmental responsibility. See Worster, *Dust Bowl*; William Boyd, "Making Meat"; Elmore, *Citizen Coke*; Cunfer, *On the Great Plains*.

19. Colin Duncan is one among several historians, philosophers, and scientists who have argued for local responsibility in land use. Some, such as Barry Commoner, did not believe that capitalism had to be hostile to nature. Others, such as Wendell Berry and Donald Worster have implied or directly stated that capitalism is always hostile to the environment. Wes Jackson, on the other hand, while largely skeptical of capitalism, believes that a scientific approach to local land use can sustain a productive and biologically healthy agriculture, even in the presence of capitalist modes of production. See Duncan, *Centrality of Agriculture*; Commoner, *Closing Circle*; Berry, *Unsettling of America*; Worster, *Dust Bowl*; Jackson, *Altars of Unhewn Stone*.

20. *Cattle Beet Capital* fits within efforts by environmental historians to analyze agroecology as a historical actor. The call to emphasize the agroecological history of specific landscapes is generally traced to an essay by Donald Worster from 1990. Books by Mort Stewart, Colin Duncan, Brian Donahue, and Geoff Cunfer directly address that call. See Worster, "Transformations of the Earth," 1087–106; Stewart, *What Nature Suffers to Groe*; Duncan, *Centrality of Agriculture*; Donahue, *Great Meadow*; Cunfer, *On the Great Plains*.

21. Nash, "Fruits of Ill-Health," 203.

22. Interpretations of the research role of land grant colleges and the USDA vary considerably. Charles Rosenberg argues that by the early twentieth century research scientists at land grant colleges and the USDA had been able to establish research prerogatives "based on shared interest in growth and productivity through the rational application of science." He also argues that, though most researchers believed in the value of small family farms, their work caused them to gravitate toward heavily capitalized farmers and industry. In her analysis of the development of hybrid corn, Deborah Fitzgerald argues

that land grants, the USDA, and the seed industry largely worked together to develop hybrid corn. However, land grant colleges, such as the University of Illinois, discontinued their support for hybrid seed research once it became clear, in the 1940s, that hybrid corn was a commercial product. In the 1973 *Hightower Report*, researchers harshly criticized land grant colleges for abandoning farmers in favor of industry. By arguing that agri-state officials failed to question how industry products harmed the land and farmers, I largely agree with the *Hightower Report*. By unfolding the historical circumstances of Hightower's agri-state critique, I go beyond his somewhat static critique. See Rosenberg, *No Other Gods*, 153–99; Fitzgerald, *Business of Breeding*; Hightower and Agribusiness Accountability Project, *Hard Tomatoes, Hard Times*.

23. Studies that conclude during the 1920s and 1930s include Fiege, *Irrigated Eden*, and Stoll, *Fruits of Natural Advantage*.

24. Andrew Duffin's work on the Palouse region of Washington State provides an exception. However, the defining environmental feature within his narrative is erosion, which played a very limited role on the Piedmont. See Duffin, *Plowed Under*.

1. CULTIVATING A REGIONAL AGROECOLOGY

1. The details of that event, as well as the relationship between Meeker and Greeley, and the early history of the Union Colony are described in David Boyd, *A History*. On Horace Greeley's farming experiments, see Greeley, *What I Know of Farming*.

2. Historian Steven Stoll uses the term intensive cultivation to describe the development of industrial agriculture in California's Central Valley during the late nineteenth and early twentieth centuries. By contrast, extensive cultivation emphasized larger acreage, minimal management, lower value crops and, typically, a willingness to mine the soil in one place before moving on to a new locale. See Stoll, *Fruits of Natural Advantage*.

3. One example of the prevalence of this idea, often described as "rain following the plow," among those who formed the Union Colony is described in Greeley, *What I Know of Farming*, 278–79.

4. As the Union Colony was the first settlement to engage in largescale irrigation on the Piedmont, its experience inspired some of the legal framework for water law in Colorado. See Schorr, *Colorado Doctrine*.

5. Wyckoff, *Creating Colorado*, 126–34.

6. Books that address the rise and fall of the open range cattle industry, as well as its promotion, include Brisbin and Newberry Library, *Beef Bonanza*; Frink, Jackson, and Spring, *When Grass Was King*; Goff, *Century in the Saddle*; von Richtofen, *Cattle-Raising*.

7. The role of early land grant experiment stations is addressed in Rosenberg, *No Other Gods*, 153–72. On the role of Colorado Agricultural College, see Steinel and Colorado Agricultural College, *History of Agriculture*, 541–81.

8. West, *Contested Plains*, 1–96.

9. According to West, Gerry's familial links to the signer of the Declaration of Independence are shaky at best. West, *Contested Plains*, 188–90, 247; Mehls, *New Empire*, 65–66; "Elbridge Gerry, Colorado Pioneer" . Descriptions of Elbridge Gerry in Colorado are also found in interviews with early Greeley settlers Walter Ennes and Eugene Williams, Civil Works Administration Pioneer Interviews (CWA), vol. 343.

10. Fite, *Farmers' Frontier*, 175–92; Wyckoff, *Creating Colorado*, 101–53; Steinel and Colorado Agricultural College, *History of Agriculture*, 47–72; West, *Contested Plains*, 97–172; Mehls, *New Empire*, 33–36.

11. On how Cheyennes used the Piedmont landscape, see West, *Contested Plains*, 71–75.

12. Sherow, *Grasslands of the United States*, 2–8; Weaver and Albertson, *Grasslands of the Great Plains*, 233–56.

13. Sherow, *Grasslands of the United States*, 8–10.

14. The literature on the movement, hunting patterns, and wars among the Plains Indians in the eighteenth and nineteenth centuries is deep. On the Sioux, see White, "Winning of the West," 319–43. On the Comanches, see Hamalainen, *Comanche Empire*. On bison numbers and their movement, see Isenberg, *Destruction of the Bison*; and Flores, *Natural West*, 50–70. On the Cheyennes, see West, *Contested Plains*, 17–96.

15. Sherow, *Grasslands of the United States*, 59–62.

16. Mehls, *New Empire*, 51–74; West, *Contested Plains*, 97–172; Sherow, *Grasslands of the United States*, 59–62.

17. Greeley and Owen, *Autobiography of Horace Greeley*, 144–58; Snay, *Horace Greeley*, 65–79; Stoll, *Larding the Lean Earth*, 204–8; Smythe, *Conquest of Arid America*, 77–91; Worster, *Rivers of Empire*, 83–96.

18. Greeley and Owen, *Autobiography of Horace Greeley*, 144–58; Snay, *Horace*, 65–79; Stoll, *Larding the Lean Earth*, 204–8; Smythe, *Conquest of Arid America*, 77–91; Worster, *Rivers of Empire*, 83–96.

19. Willard, *Union Colony at Greeley*, xix–xxv; Wyckoff, *Creating Colorado*, 126–30; Boyd, *A History*, 29–38; "Meeker, Nathan Cook," *National Cyclopedia of American Biography* 8 (January 1898): 387; Decker, "*Utes Must Go!*," 69–90.

20. Historian Conevery Bolt Valenčius argues that how settlers perceived the health of land in the nineteenth century helped to determine whether they settled and stayed there. See Valenčius, *Health of the Country*.

21. Nathan Meeker, "A Western Colony," *New York City Tribune*, December 4, 1869.

22. Meeker, "Western Colony"; Nathan Meeker, "Colonization: The Organization of the Colorado Colony," *New York Tribune*, December 24, 1869.

23. Willard, *Union Colony at Greeley*, xix–xxxi; Wyckoff, *Creating Colorado*, 126–34; Boyd, *A History*, 29–63.

24. "Greeley," *Denver Tribune*, April 13, 1871, 4.

25. Etta Kettley, Mary Norcross Tuckerman, Mrs. Pitt-Smith, and Jennie Lucas Interviews, CWA Interviews, vol. 343; Union Colony of Colorado, *First Annual Report*.

26. Piedmont settler Asa McLeod states that herds of bison were completely gone from the plains of Northern Colorado by 1880. See Asa McCloud interview, CWA Interviews, vol. 343; see also Norcross, Pitt-Smith, and Lucas Interviews, CWA Interviews, vol. 343. On the decimation of bison, see Isenberg, *Destruction of the Bison*.

27. "Land Speculation," *Greeley Tribune*, April 5, 1871, 2; "Territorial News," *Rocky Mountain News*, June 1, 1871, 1; Clark, *Colonial Days*, 138.

28. Baker, "Recollections of the Union Colony," 145.

29. To discourage subdividing lots, Greeley imposed a five dollar tax on each lot division. See Boyd, *A History*, 75–87.

30. According to Boyd, the total cost of building Ditches Number Two and Three was $412,000; see Boyd, *A History*, 45–63. See also Laflin, *Irrigation, Settlement, and Change*, 16–22; Worster, *Rivers of Empire*, 83–96.

31. Boyd, *A History*, 59–63, and Laflin, *Irrigation, Settlement, and Change*, 16–22. For other examples of how these early attempts to build irrigation infrastructure struggled with nature and lack of knowledge, see Fiege, *Irrigated Eden*, 3–80 and Morison, *From Know-How to Nowhere*, 40–71.

32. Maass and Anderson, *And the Desert Shall Rejoice*, 290–91; Boyd, *A History*, 119–21; Worster, *Rivers of Empire*, 87.

33. Worster, *Rivers of Empire*, 83–96.

34. Irrigation, municipal, and industrial uses were generally considered "beneficial use."

35. Laflin, *Irrigation, Settlement, and Change*, 41–53; Schorr, *Colorado Doctrine*; Worster, *Rivers of Empire*, 83–96.

36. Hutchins, *Mutual Irrigation Companies*; Mead, *Irrigation Institutions*, 143–79. While historian David Schorr argues that these mutual irrigation companies were largely practical responses to aridity, Donald Worster argues that mutual irrigation companies were formed to exploit nature for profit. See Schorr, *Colorado Doctrine* and Worster, *Rivers of Empire*, 83–96.

37. Hutchins, *Mutual Irrigation Companies*; Mead, *Irrigation Institutions*, 143–79.

38. Willard, *Experiments in Colorado Colonization*, 1–26.

39. Steinel and Colorado Agricultural College, *History of Agriculture*, 167–244; Wyckoff, *Creating Colorado*, 126–33; Hemphill and United States, eds., *Irrigation in Northern Colorado*.

40. Mead, *Irrigation Institutions*, 172.

41. The process by which nature became capital is examined thoroughly in Cronon, *Nature's Metropolis*.

42. The history of cattle ranching on the Great Plains has been told several times. Histories tend to highlight the entrepreneurial skills and individual initiative of ranchers, including Webb, *Great Plains*; Frink, Jackson, and Spring, *When Grass Was King*; Goff, *Century in the Saddle*; Dale, *Range Cattle Industry*. For a more recent interpretation that analyzes how Great Plains cattle ranching impacted regional ecosystems and how it was integrated with industrial capitalism see Specht, *Red Meat Republic*, 1–118.

43. Goff, *Century in the Saddle*, 14–16.

44. Goff, *Century in the Saddle*, 4–6; Steuter and Hidinger, "Comparative Ecology of Bison," 329–42.

45. Steuter and Hidinger, "Comparative Ecology of Bison," 329–42; Cronon, *Nature's Metropolis*, 213–24.

46. Jenkins, "Hardworking John Wesley Iliff."

47. Isenberg, *Destruction of the Bison*, 123–63; Agnes Wright Spring asserts that Iliff's cattle were occasionally rustled but that this had little impact on his holdings; see Frink, Jackson, and Spring, *When Grass Was King*, 345–65. While most Piedmont settlers indicate either friendly or minimal interaction with Indians, some describe tensions and conflict. Examples include A. D. Bennett, Walter Ennes, and Grace Brush Mayne, cwa Interviews, vol. 343.

48. Steinel and Colorado Agricultural College, *History of Agriculture*, 107–65.

49. Spring in Frink, Jackson, and Spring, *When Grass Was King*, 366–82.

50. Spring in Frink, Jackson, and Spring, *When Grass Was King*, 334–441; John Wesley Iliff is quoted in Bray, *Financing the Western Cattleman*, 12.

51. Examples of books that oversold plains cattle ranching include Brisbin, *Beef Bonanza*; von Richthofen, *Cattle-Raising*.

52. Frink, Jackson, and Spring, *When Grass Was King*, 96–102; Goff, *Century in the Saddle*, 121–32; Dale, *Range Cattle Industry*, 94–101.

53. Sherow, *Grasslands of the United States*, 8–9 and 57–63; Cronon, *Nature's Metropolis*, 218–24; Specht, *Red Meat Republic*, 106–18; Bray, *Financing the Western Cattleman*.

54. Wyckoff, *Creating Colorado*, 160–61; Steinel and Colorado Agricultural College, *History of Agriculture*, 107–65; Dale, *Range Cattle Industry*, 77–101; Frink, Jackson, and Spring, *When Grass Was King*, 93–123; Mehls, *New Empire*, 65–74.

55. The first available journal for Philip Boothroyd was in the year 1872. I read his journals for the following years: 1872, 1893, 1899, 1906, 1914, 1919, and 1929. See Philip H. Boothroyd, Manuscript Collections.

56. Boothroyd, 1872.

57. Boothroyd, 1893 and 1899. Boothroyd also joined the American Aberdeen-Angus Breeders' Association no later than 1893.

58. For an example of a Colorado Agricultural College study on the feed value of corn, alfalfa, oats, and beets, see Cooke, *Cattle Feeding in Colorado*, 3–36.

59. To examine how cattle ranchers came to see the need for water, I analyzed the articles of incorporation for eighteen cattle ranching operations that were dated before 1887. Only two of them applied for water rights. Of the articles after 1887, two-thirds applied for water rights. Selected articles of incorporation were examined from the *Western Range Cattle Industry Study*, manuscript study 699, Stephen H. Hart Library.

60. On the establishment of land grant colleges and experiment stations see Rosenberg, *No Other Gods*, 153–72; Rossiter, "Organization of the Agricultural Sciences," 211–48; For a larger discussion of the philosophy behind the land grant colleges see Eddy, *Colleges for Our Land and Time*, 1–112; For details on how the Colorado Agricultural College and its experiment station were established, see; Steinel and Colorado Agricultural College, *History of Agriculture*, 541–81.

61. Alan Marcus describes a tension among farmers regarding the degree to which they trusted academia to supply agronomic science relevant to their farming practices and the degree to which they were willing to relinquish agricultural knowledge production to university-trained professionals. On the Piedmont, this tension appears to be subdued as farmers were eager to apply modern science to farming practice. Moreover, experiment station researchers at CAC (Colorado Agricultural College) focused most of their research on science that had direct application to Colorado farmers. On the aforementioned tension, see Marcus, *Agricultural Science*, 7–26, 188–220.

62. On the development of agricultural sciences in the nineteenth century, see Rossiter, *Emergence of Agricultural Science*, 10–28 and 126–76.

63. Boyd, *A History*, 142–55.

64. Rasmussen, *Readings in the History*, 94–102; Olmstead, *Creating Abundance*, 277–78.

65. Olmstead, *Creating Abundance*, 277–78; Mehls, *Weld County*, 7–8. While a report from the Experiment Station at CAC states that farmers planted alfalfa in the South Platte Valley in the 1860s, fragmentary evidence from the *Colorado Farmer* makes no mention of alfalfa in its earliest available editions in 1874. However, in an article devoted to the crops grown in the Greeley area in 1879, the author describes several farmers who devoted fields to alfalfa. See Colorado Agricultural Experiment Station, *Alfalfa*; and "Weld County, Greeley and the Union Colony."

66. Boyd, *A History*, 145.

67. Colorado Agricultural Experiment Station, *Alfalfa*; Boothroyd, 1893 diary.

68. Wyckoff, *Creating Colorado*, 101–3; Steinel and Colorado Agricultural College, *History of Agriculture*, 11–90.

69. For an example of corporate farming in California in 1900, see Stoll, *Fruits of Natural Advantage*.

70. Wyckoff, *Creating Colorado*, 124–33.

71. Snay, *Horace Greeley*, 176–82; Decker, *"Utes Must Go!,"* 121–44.

2. CAPITALISM AND FARMING

1. May, *Great Western Sugarlands*, 1–28; Bean, *Charles Boettcher*.

2. Though chemists demonstrated that cane and beet sugar were identical long before Great Western Sugar arrived on the Piedmont, the company was careful to make this plain to consumers. See, for example, C. A. Browne, acting chief, USDA Dept. of Chemistry to Great Western Sugar, October 12, 1923, reprinted in *Through the Leaves* (TTL), April 1924.

3. The characteristics of capitalism described in this paragraph were adapted from Walker, *Conquest of Bread*, 1–18.

4. Deborah Fitzgerald identifies the following characteristics of industrial agriculture: assembly lines in harvesting and processing, orientation toward the clock, the presence of farm unions, mass production, operation timelines, largescale production sites, mechanization, product standardization, on-farm specialization, speed of throughput, a de-skilled workforce, and an overriding notion of efficiency in all tasks. She identifies the period, 1918–31 as the most critical for developing this mentality in farming. In their analyses of California agriculture Steven Stoll and Richard Walker argue that farmers oriented their production around capitalism and industrial production from the start. In fact, Stoll refers to the subjects of his study as orchard capitalists rather than farmers. Seeking to understand the trajectory of modern agriculture from the perspective of the land, Colin Duncan argues that the goal of capitalism in the fields is to simplify ecological relationships so that nature

can be made to operate according to industrial prerogatives. See Fitzgerald, *Every Farm a Factory*, 1–32; Walker, *Conquest of Bread*, 1–75; Stoll, *Fruits of Natural Advantage*; Duncan, *Centrality of Agriculture*, 3–49.

5. I refer to farmers who cultivated sugar beets for Great Western as growers. In this sense, I define the term more generically than other historians who view growers within the context of horticulture, especially that which developed in the California citrus industry during the late nineteenth and early twentieth centuries. To varying degrees Richard Sawyer, Steven Stoll, and Douglas Sackman emphasize that growers in the citrus industry viewed themselves less as farmers and more as entrepreneurs and educated professionals, or "orchard capitalists," as Stoll calls them. They emphasized irrigation, intensive agriculture on small acreage, cooperative marketing, single-species, the use of both biological and chemical control, and cozy relationships with the agri-state scientists who performed relevant research. While most Piedmont farmers grew beets for Great Western and embraced agricultural science, sugar beet agroecology, climate, and the regional history of land ownership and tenure all precluded them from fitting into the California citrus industry's grower model. See Sawyer, *To Make a Spotless Orange*, 17–37; Stoll, *Fruits of Natural Advantage*, 30–62; Sackman, *Orange Empire*, 20–52.

6. This is not to say that farmers did not purchase fertilizers. However, studies performed by Great Western Sugar, the USDA, and agronomists at CAC into the 1920s consistently showed that they provided little or no benefit over prevailing practices. See, for example, Moorhouse et al., *Farm Practice*.

7. In comparable studies, Colin Duncan and Brian Donahue found that farming in Victorian England and eighteenth-century Massachusetts were relatively sustainable as measured by carefully calibrated soil management and forms of animal husbandry that managed animals according to how much manure they could provide. See Duncan, *Centrality of Agriculture*, 63–71; Donahue, *Great Meadow*.

8. While there is slight variation in statistics, it is clear that by 1914, the Piedmont region was the nation's largest domestic producer. See Sabin, *Colorado Beet Boom*; May, *Great Western Sugarlands*, 61–71.

9. Anderson "Geography of the Sugar Beet Industry," 16–25; Don Ament, cattle feeder and former Colorado Agricultural Commissioner, interview by author, Iliff, June 23, 2019.

10. Mintz, *Sweetness and Power*, 19–150; Coons, "Sugar Beet," 149–64.

11. Ellis et al., *Tariff on Sugar*, 45.

12. On arguments in support of the beet sugar industry see Merleaux, *Sugar and Civilization*, 28–54.

13. Merleaux, *Sugar and Civilization*, 28–54; Mapes, *Sweet Tyranny*, 1–12; Saylor, *Progress of the Beet-Sugar Industry*, 46.

14. On the tactics used by the American Sugar Refining Company, see Zerbe, "American Sugar Refinery Company," 339–75.

15. Among historians who examine the role of economic protectionism in jump-starting and maintaining the beet sugar industry, only Leonard Arrington argues that the tariff was of little consequence. Arrington claims that the role played by various beet sugar entrepreneurs was of much greater value. By contrast, April Merleaux, Kathleen Mapes, and William John May Jr. agree that high tariffs were essential. Given that these same entrepreneurs were some of the strongest lobbyists for protection and that the industry's rise was so clearly timed with high tariffs, overwhelming evidence supports Merleaux, Mapes, and May. See Arrington, "Science, Government, and Enterprise," 1–18; Merleaux, *Sugar and Civilization*, 28–54; Mapes, *Sweet Tyranny*, 1–16; May, *Great Western Sugarlands*, 220–37.

16. According to historian Kristin Hoganson, one additional factor in supporting Cuba had to do with notions of manliness that came into play, as many American males felt it was their duty to fight against perceived atrocities against Cuban women and thus uphold the dignity of the American male. See Hoganson, *Fighting for American Manhood*, 43–67.

17. Merleaux, *Sugar and Civilization*, 28–54; Mapes, *Sweet Tyranny*, 16–34.

18. Mapes, *Sweet Tyranny*, 26–34.

19. Some examples of beet research and promotion can be found in Colorado Agricultural Bulletins 7 (1889), 11 (1890), 14 (1891), 23 (1893), 43 (1898), and 51 (1899). On how the beet sugar industry promoted itself in general, see Saylor, *Progress of the Beet-Sugar Industry*, 11.

20. Saylor, *Progress of the Beet-Sugar Industry*, 27–42; Sabin, *Colorado Beet Boom*, 19–20.

21. Sabin, *Colorado Beet Boom*, 20–21; May, *Great Western Sugarlands*, 44–60. On Havemeyer's reputation as a brutal taskmaster and the cutthroat methods he employed to destroy competition see Elmore, *Citizen Coke*, 80–82.

22. Bean, *Charles Boettcher*, 44–60.

23. "Water Supply of Sugar Factory," *Sugar Press*, February 5, 1920, 8.

24. On conflicts over water, see for example H. N. Haynes to A. V. Officer, December 9, 1903, GWS Loveland, box 2. On conflicts over coal, see for example Colorado and Southern Railway to A. V. Officer, December 11, 1903, GWS Loveland, box 1.

25. May, *Great Western Sugarlands*, 61–68.

26. While the Piedmont was Great Western's center of production, the company also owned factories in Wyoming and Nebraska. The beet sugar industry also developed a significant presence in Michigan, California, and Utah, as well as more modest operations in other western states and the Midwest. See Townsend, *Beet-Sugar Industry*; Bean, *Charles Boettcher*, 166–70; May, *Great Western Sugarlands*, 61–71.

27. Jessen, *Railroads of Northern Colorado*, 69–131; Morgan, *Sugar Tramp*, 5–20.

28. Jessen, *Railroads of Northern Colorado*, 177.

29. Charles Boettcher to Colorado Fuel and Iron, n.d., GWS Loveland, box 2.

30. "Water Supply of Sugar Factory."

31. The subject of water and industrial agriculture in the region will be addressed in depth in chapter 5. For a brief treatment of Great Western and water, see May, *Great Western Sugarlands*, 203–19.

32. May, *Great Western Sugarlands*, 44–71.

33. Several historians have examined the relationship between capitalism and agroecology in the early twentieth-century American West. Richard Walker, Steven Stoll, and Alan Olmstead argue that exploiting land, labor, and soil for wealth accumulation occupied a central position in California at the time. Stoll argues that California fruit growers did not see themselves as farmers, but as 'orchard capitalists' who oriented land, labor, machines, chemicals, and state-sponsored scientists around intensive production of high value crops. Walker emphasizes that the structures of capitalism did not evolve in response to agricultural development in California but were present when Anglos arrived after 1850. Olmstead largely agrees but focuses on how California farmers embraced the monocropping of high-value exotic cultivars and rapidly adopted chemical pesticides to protect their investments. As Andrew Duffin shows, the monocropping of wheat on the Palouse in Eastern Washington was already well developed by 1900, as farmers sunk all their resources into mining the region's soils. Emphasizing the unpredictable world made by irrigation on Idaho's Snake River Plain, Mark Fiege shows how farmers there sought to transform their farms into discrete units of capitalism but were consistently thwarted by nature's failure to act according to their designs. Deborah Fitzgerald shows how, during the 1920s, Thomas Campbell used factory-based principles and economies of scale to manage a two-hundred-thousand-acre farm in Montana. Though none of the above are examples of one corporation monopolizing production of a single crop throughout a region, they all demonstrate efforts to use the tools of capitalism to control the land and labor of entire regions. See Stoll, *Fruits of Natural Advantage*;

Walker, *Conquest of Bread*; Olmstead, *Creating Abundance*, 223–61; Duffin, *Plowed Under*; Fiege, *Irrigated Eden*; Fitzgerald, *Every Farm a Factory*, 129–56.

34. Books that follow this trajectory of power concentration during the late nineteenth and early twentieth centuries, albeit on different subjects, include Walker, *Conquest of Bread*; Richard White, *Railroaded*; Worster, *Rivers of Empire*.

35. Experiments conducted by CAC in the early twentieth century with the use of commercial nitrogen, phosphates, and potassium had little impact on productivity. The only chemical of note used by Piedmont farmers prior to 1930 was Paris Green, and this was employed only during periodic insect infestations. See, for example, C. O. Townsend to Sugar Beet Companies, September 6, 1904, SPI, Office Files of pathologist C. O. Townsend, box 4, RG 54; A. C. Maxson, "The Beet Webworm," *TTL*, June 1913.

36. The only exception to this is corn. Those feeding beef cattle for the market typically imported corn from the Midwest since it produced weight gain more rapidly than any other grain grown in quantity.

37. While the economic depression in Colorado that began in 1893 created unstable conditions on the Piedmont, especially in livestock production, it did not alter Piedmont land use dramatically. While it is true that the depression increased the appeal of the beet sugar industry, I argue that the industry solidified existing practices through the introduction of a new crop. On Colorado's economy during the 1890s, see Ubbelohde, Benson, and Smith, *A Colorado History*, 195–249.

38. On Piedmont crop rotations and irrigation in 1900 see Steinel and Colorado Agricultural College, *History of Agriculture*, 167–244; Watrous, "Sugar Beets"; Boyd, *A History*, 142–55; Wyckoff, *Creating Colorado*, 160–61.

39. *Colorado Agricultural Bulletins* 7 (1889), 11 (1890), 14 (1891), 23 (1893), 43 (1898), and 51 (1899); Saylor, *Progress of the Beet-Sugar Industry*. There are many examples of correspondence between the USDA's Division of Sugar Plant Investigations and Great Western Sugar regarding effective crop rotations and soil management. See, for example "Summary of Letters Received in Relations to Crop Rotation in Beet Districts, 1904–09," SPI, Townsend Office Files, box 1, "Crop Rotation," RG 54.

40. Faurot, *Practical Talk*; Moorhouse et al., *Farm Practice*; O. E. Baker, *Graphic Summary of American Agriculture*; United States Tariff Commission, *Sugar*.

41. On the origins and trajectory of livestock feeding in Iowa and Illinois, see Whitaker, *Feedlot Empire*.

42. Henry and Morrison, *Feeds and Feeding*, 345–64; Colorado Agricultural Experiment Station, *Alfalfa*; Cooke, *Cattle Feeding in Colorado*.

43. As categories of livestock feed, alfalfa is a succulent and corn kernels constitute a grain, and both are needed in the feeding regimen. Wheat and barley were the two grains most commonly fed on the Piedmont, and feed returns on those grains could not approach those afforded by corn. The high land costs of irrigated farms also created a barrier for Piedmont farmers. Corn could not yield returns high enough to justify farmer expenses, especially since the majority of farmers either carried mortgages or were tenant farmers.

44. Cooke and Headden, *Sugar Beets in Colorado*; Spencer, *Utilization of Residues*; A. H. Heldt, "Value of Beet Tops to Growers," *TTL*, October 1924, 512.

45. Anderson, "Geography of the Sugar Beet Industry," 93–101; John Grant, "Feeding Experiments," *TTL*, 1918, 121 and 153.

46. *TTL*, December 1913, 9.

47. Examples of Maynard's research published in journals from CAC's Experiment Station include Maynard, *Beet By-Products for Fattening Lambs*; Maynard, *Feedlot Fattening*; Maynard, *Colorado Fattening Rations*. Examples of Maynard's cattle feeding research published in *Through the Leaves* include "Value of Beet Tops When Fed to Steers," *TTL*, February 1920; "Value of Dried Beet Pulp," *TTL*, September 1924; "Fattening Beef Calves in Colorado," *TTL*, 1927, 74.

48. Maynard, *Beets and Meat*, 11–34.

49. Between 1890 and 1920, U.S. beef consumption per capita declined from eighty-eight pounds to fifty-seven pounds. See Clemen, *American Livestock and Meat Industry*, 255–57; Bussing, "Cattle Feeding Industry," 33–73. For one example where Maynard encourages growers to feed cattle despite losses, see Maynard, *Feedlot Fattening Rations*.

50. Historian Brian Donahue advances a similar argument about the interdependencies of farmers and the land for the Concord, Massachusetts, region during the 1700s. In the Concord example, farmers prioritized stable, diversified farming in the face of market demands. See Donahue, *Great Meadow*. Even in cases where extra-local fertilizer was used, studies showed little or no increased soil fertility. See Moorhouse et al., *Farm Practice*.

51. According to Charles Townsend, head of the USDA's Division of Sugar Plant Investigations, a typical beet sugar factory in the United States required 5,700 acres of sugar beets annually to operate at full capacity. See C. O. Townsend, "Conditions Affecting," *Yearbook of the U.S. Department of Agriculture, 1909* (Washington: GPO, 1910), 174.

52. Moorhouse et al., *Farm Practice*.

53. May, *Great Western Sugarlands*, 72–114.

54. There are many examples of Great Western's concerns over losing its hold on the sugar market in the plains and the Midwest in its company publications. See for example *TTL*, April 1924, 208; *TTL*, November 1929, 520–22; "Beet Sugar," *Iowa Homestead*, June 5, 1924, reprinted in *TTL*, June 1924, 385.

55. Twitty, *Silver Wedge*, 45.

56. Hans Mendelson, "You and Your Sugar Beet Industry," *TTL*, January 1919, 8–19.

57. Lowell Giaque, interview by Novella Myers, December 8, 1999, Loveland Historical Society Museum; Norman Vlass, interview by Novella Myers, January 5, 2000, Loveland Historical Society Museum.

58. On the various aspects of capitalist agriculture, see Walker, *Conquest of Bread*, 1–18.

59. Walker, *Conquest of Bread*, 1–18.

3. BIOLOGY AND LABOR

1. Kloppenburg, *First the Seed*, 1–18.

2. For a larger discussion of the social, economic, and ecological consequences attached to the choice of which plants to grow, see Kloppenburg, *First the Seed*, 1–18. On the nature of beet seeds and the labor requirements that attended them, see "Thinning Beets," *TTL*, May 1920, 280–82.

3. Since German Russians, Hispanos, and Mexican nationals provided, by far, the largest source of labor for GWS, this chapter focuses primarily on these groups. Hispano laborers who worked in the beet fields of the Piedmont came primarily from northern New Mexico and southern Colorado and could trace their presence in those regions to the period before the Treaty of Guadalupe-Hidalgo in 1848. By referring to themselves as Hispanos, they were both embracing their Spanish heritage and distinguishing themselves from laborers imported from Mexico. In studies of the Southwest, Katherine Benton-Cohen, Neil Foley, and David Gutiérrez have shown that the referent "Hispano" was common in New Mexico and Colorado, but less common in Texas, Arizona, and California. In cases where it is clear that my sources refer to laborers imported from Mexico, I will use the term Mexican national(s). However, many sources simply refer to Mexicans because Great Western Sugar and most Piedmont residents made no distinction between Mexican nationals and Hispanos. Where no distinction is clear, I use the term "Mexican," since it is the term used by my sources. On variations of nomenclature see Deutsch, *No Separate Refuge*; Benton-Cohen, *Borderline Americans*; Foley, *White Scourge*; Gutiérrez, *Walls and Mirrors*.

4. *Colorado: Report on the Farmers' Costs.*

5. While many reports emerged beginning in 1916 on the plight of contract laborers in the sugar beet fields, four focus on child labor. See Clopper and Hine, *Child Labor in the Sugar-Beet Fields*; Matthews, Light, and United States, *Child Labor and the Work of Mothers*; Coen, *Children Working on Farms*; Johnson, *Welfare of Families*.

6. Much of the scholarly attention paid to agroecology within environmental history can be traced to a 1990 essay by Donald Worster in which he argued that historians should pay attention to how capitalist modes of production restructure ecosystems. Taking cues from Worster, Mort Stewart and Brian Soluri both examine how humans simplify landscapes to facilitate certain kinds of agricultural production to synchronize with the global capitalist economy. In their work on agricultural laborers, Linda Nash and Brian Sackman conclude that workers were deeply embedded within the cultivation processes in which they labored. I extend these arguments by arguing that sugar beet laborers were as embedded in its agroecology as the various organisms that made it possible. See Worster, "Transformations of the Earth," 1087–106; Stewart, *"What Nature Suffers to Groe"*; Soluri, *Banana Cultures*; Nash, "Fruits of Ill-Health," 203; Sackman, "'Nature's Workshop,'" 27–53.

7. Several historians have examined the social, cultural, and economic relationships of sugar beet laborers. However, none have explicitly examined how environmental risk was structured into worker contracts. Sarah Deutsch examines the formation and development of Hispano communities of contract laborers in Northern Colorado. In a chapter on "beeters," Mark Wyman provides an overview of Great Western's efforts to attract marginalized laborers to the Piedmont, with some emphasis on the experiences of laborers in the fields. Kathleen Mapes and April Merleaux both connect the experience of contract laborers with larger questions of immigration policy, race, and global sugar production. See Deutsch, *No Separate Refuge*; Wyman, *Hoboes*, 170–98; Mapes, *Sweet Tyranny*; Merleaux, *Sugar and Civilization*.

8. Saylor, *Progress of the Beet-Sugar Industry in the United States in 1901*, United States Congressional Serial Set, no. 4241 (Washington, GPO, 1902); Coons, "Sugar Beet," 149–64.

9. Cooke and Headden, *Sugar Beets in Colorado*, 29.

10. Examples that show this knowledge of the labor needs of the beet sugar industry can be found in Dupree, *Science in the Federal Government*, 176–80; Colorado Agricultural Bulletins 7 (1889), 11 (1890), 14 (1891), 23 (1893), 43 (1898), and 51 (1899); *Ranch and Range*, 1902, 5–33.

11. For examples, see Andrews, *Killing for Coal*, 87–121; Stoll, *Fruits of Natural Advantage*, 124–54; Sackman, "Nature's Workshop," 27–53.
12. For a thorough account of the experience of Germans in Russia, including their reasons for immigration to the Americas, see Kloberdanz, *Volga Germans in Old Russia*; see also Williams, "Social Study of the Russian German."
13. Shaw, "Twice Separated," 21–40.
14. Shaw, "Twice Separated," 41–61.
15. Much of the literature on this subject centers around the degree to which immigrants from Eastern Europe were deemed as fit for assimilation into American society and could adopt the tenets of democracy. Works on this subject include Higham, *Strangers in the Land*; and Jacobson, *Whiteness*.
16. Williams, "Social Study of the Russian German"; Wyman, *Hoboes*, 170–98; Standish, "Beet Borderland."
17. Wyman, *Hoboes*, 170–98.
18. Shaw, "Twice Separated," 62–81.
19. Shaw, "Twice Separated," 41–81; Wyman, *Hoboes*, 170–98; Haynes, "German-Russians on the Volga," 118–70.
20. Rock, "'Unsere Leute,'" *Colorado Magazine*, Spring 1977, 154–83.
21. On the worldwide sugar shortage during World War I, see Dalton, *Sugar*, 19–39.
22. On prices paid per acre over time, see Schwartz, *Seasonal Farm Labor*, 128.
23. Taylor, *Mexican Labor in the United States*, 103–4.
24. C. V. Maddux, "Beet Labor at $20.53," *TTL*, August 1919, 378–79; Deutsch, *No Separate Refuge*, 107–26.
25. Higham, *Strangers in the Land*, 158–263; Jacobson, *Whiteness*, 39–90.
26. Shaw, "Twice Separated," 82–104.
27. Koch, *Volga Germans*, 212–21.
28. Gonzalez, "Mexican Labor Migration."
29. Deutsch, *No Separate Refuge*, 117–26.
30. U.S. Department of Labor, *Report of the Special Committee*.
31. U.S. Congress, House, *Seasonal Agricultural Laborers from Mexico*, 66.
32. U.S. Congress, House Committee on Immigration and Naturalization, *Immigration from Countries*, 246.
33. Taylor, *Mexican Labor*, 196, 150–62, 234.
34. Wyman, *Hoboes*, 170–98.
35. J. F. Jarrell, "Land Values," *TTL*, 1919, 90–91.
36. Taylor, *Mexican Labor*, 176–84.

37. Deutsch, *No Separate Refuge*, 107–61. For an example of the discrimination suffered by Hispanos working in the mines of the Southwest during World War I, see Benton-Cohen, *Borderline Americans*.

38. Schwartz, *Seasonal Farm Labor*.

39. While each beet sugar producing region of the United States was distinct, they all employed contract migrant laborers to complete identical thinning, cultivation, and harvesting tasks. Substandard living conditions, social segregation, and educational marginalization were all features of the beet sugar industry. Consequently, while much of this section draws on the experiences of Piedmont beet laborers, I also employ oral histories from other sugar beet producing regions such as Minnesota, Wisconsin, and Michigan.

40. On the mechanization of American agriculture in the early twentieth century, see Conkin, *Revolution down on the Farm*, 1–30; and Fitzgerald, *Every Farm a Factory*, 9–32. On the slow mechanization of sugar beet cultivation, see Rasmussen, "Technological Change," 31–36. On sugar beet laborers, see Wyman, *Hoboes*, 170–98. In her examination of a community of beetworkers, Standish argues that "the nature of the sugar beet determined the nature of the labor." See Standish, "Beet Borderland," 42.

41. Ham, "Regulation of Labour Conditions," 74–82.

42. Elva Treviño Hart, *Barefoot Heart*, 33–50; Lopez, Lopez, and Ford, *White Gold Laborers*, 40–61.

43. Taylor, *Mexican Labor*, 128.

44. Anderson, "Geography of the Beet Sugar Industry," 16–42.

45. Minnesota beet laborers could find work in the onion and lettuce fields of Wisconsin, while California workers had several options, including fruit and nut orchards.

46. Rasmussen, "Technological Change." Beet toppers existed prior to the 1930s but were not in common use since they eliminated too much of the sugar-rich portion of the beet.

47. Gilbert Barela, former sugar beet laborer, interview by author, Greeley CO, June 20, 2019; Sánchez, *Rows of Memory*, 73–75; Lopez, Lopez, and Ford, *White Gold Laborers*, 44–46.

48. Ham, *Regulation of Labour Conditions*.

49. Matthews, Light, and United States, *Child Labor and the Work of Mothers*, 31; Sánchez, *Rows of Memory*, 14–22; Hart, *Barefoot Heart*, 119–31; Lopez, Lopez, and Ford, *White Gold Laborers*, 13–31; Johnson, *Welfare of Families*.

50. Samples of publications include Clopper and Hine, *Child Labor in the Sugar-Beet Fields*; Matthews, *Child Labor and the Work of Mothers*; Taylor, *Mexican Labor*; Coen, *Children Working on Farms*; Johnson, *Welfare of Families*.

51. Matthews, Light, and United States, *Child Labor and the Work of Mothers*.

52. Matthews, Light, and United States, *Child Labor and the Work of Mothers*, 24–31; Standish, "Beet Borderland," 49–50; Hart, *Barefoot Heart*, 33–50; Hernandez and Hernandez-Chavez, *Elvira*, 29–58.

53. Lindenmeyer, *Right to Childhood*, 108–13, 135–38.

54. Clopper and Hine, *Child Labor in the Sugar-Beet Fields*; Wyman, *Hoboes*, 170–98; Matthews, Light, and United States, *Child Labor and the Work of Mothers*; Hernandez and Hernandez-Chavez, *Elvira*, 29–58; Mahony, *Problem of the Mexican Wage Earner*.

55. Taylor, *Mexican Labor*, 192–207; Lopez, Lopez, and Ford, *White Gold Laborers*, 136–37; Matthews, Light, and United States, *Child Labor and the Work of Mothers*.

56. Donato, *Mexicans and Hispanos*, 13–28; Coen, *Children Working on Farms*; Matthews, Light, and United States, *Child Labor and the Work of Mothers*; Standish, "Beet Borderland," 66. By the late 1920s, *colonias* existed in Fort Collins, Brush, Kersey, Johnstown, Hudson, Orchard, Ovid, Sedgewick, and in Nebraska.

57. Sánchez, *Rows of Memory*, 84, 166–67; Matthews, Light, and United States, *Child Labor and the Work of Mothers*; Hart, *Barefoot Heart*, 7–9; Coen, *Children Working on Farms*, 93.

58. Taylor, *Mexican Labor*, 176–84.

59. Hart, *Barefoot Heart*, 43; Frijoles guisados is a dish of stewed beans. In the case of beetworkers, the beans would be stewed with whatever was available in their food rations, tomatoes, peppers, and sometimes salted pork.

60. Standish, "Beet Borderland," 45–60; Sánchez, *Rows of Memory*, 22; Hernandez, *Elvira*, 29–58; Deutsch, *No Separate Refuge*, 127–61.

61. For an example of how the labor contract resulted in hiring temporary labor that reduced the wages of contract workers see E. Ward Jr., "The Use of Town Boys in Thinning and Hoeing Beets," *TTL*, August 1924.

62. Deutsch, *No Separate Refuge*, 162–99; "Minimum Wages for Sugar Beet and Sugar Cane Labor," *Monthly Labor Review*, July 1941, 169; C. V. Maddux, "Beet Field Labor for 1929," *TTL*, March 1929, 138.

63. Vargas, *Labor Rights*, 70–76.

64. "Mexican Needs More than Praise for the Expense He Is Put through in Unusual Labor in the Beets this Year," *Denver Catholic Register*, November 18, 1929; Thomas Mahony to Linna K. Bresette, Field Secretary of National Catholic Welfare Council, January 6, 1930, box 1, folder 10.

65. "The Beetworkers Association"—lists of unpaid labor in Ft. Lupton in 1929 and 1930 which were filed for claims (assumes that many did not file claims)—

TFM, box 3, folder 3; Mahony to Maddux, May 1930, TFM, box 1, folder 10; Mahony to Bresette.

4. SUGAR AND SCIENCE

1. Dupree, *Science in the Federal Government*, 176–80. Additional examples of beet research and promotion include *Colorado Agricultural Bulletins 7* (1889), 11 (1890), 14 (1891), 23 (1893), and 43 (1898).
2. The USDA's Department of Chemistry and several agricultural colleges conducted beet sugar research prior to 1903. See Dupree, *Science in the Federal Government*, 176–80.
3. On the willingness of the federal government to assert its prerogatives in the West, see McGerr, "Is There a Twentieth Century West?"; and White, *"It's Your Misfortune and None of My Own."* On how federal capital and expertise defined the relationship between the federal government and the West, see Worster, *Rivers of Empire*; Karen Merrill summarizes each of these historians in "In Search of the 'Federal Presence.'"
4. On the development of research agendas in land grant colleges at the turn of the twentieth century, see Rosenberg, *No Other Gods*, 153–99; and Rossiter, "Organization of the Agricultural Sciences," 211–48.
5. For an overview of the debates occurring within the USDA and land grant colleges about their relationship with farmers in the early twentieth century, see Fitzgerald, *Every Farm a Factory*, 29–74.
6. Rasmussen and Baker, *Department of Agriculture*; Sanders, *Roots of Reform*, 391–94; Dupree, *Science in the Federal Government*, 157–83; Carpenter, *Forging of Bureaucratic Autonomy*, 212–54.
7. Rosenberg, *No Other Gods*, 153–99.
8. At the start of the twentieth century, the most significant concentration of beet sugar factories were in Michigan, but these were quickly eclipsed by those in the West. On the beet sugar industry in Michigan, see Mapes, *Sweet Tyranny*. Institutional histories of beet sugar companies in the West include Arrington, *Beet Sugar in the West*; May, *Great Western Sugarlands*; Magnuson, "History of the Beet Sugar Industry," 68–79.
9. The USDA's longstanding emphasis on seed development and distribution is explained in Dupree, *Science in the Federal Government*, 157–83.
10. Charles Townsend to Amalgamated Sugar, January 6, 1905, SPI, Townsend Office Files, box 1, RG 54; Townsend, "Testing Comparative Merits of American and Foreign Grown Varieties of Sugar Beet Seed," Miscellaneous Reports, 1903–1909, SPI, box 1, RG 54.

11. Statement of Beets Grown by William Stanley for Great Western Sugar, 1902–1911, SPI, Townsend Office Files, box 1, RG 54.

12. W. A. Dixon to Townsend, November 16, 1904, SPI, Townsend Office Files, box 1, RG 54.

13. "Breaking and Establishing High Grade Pedigree Strains of Sugar Beet Seed," 1907, and "Commercial Production of Sugar Beet Seed," November 1, 1910, SPI, Townsend Office Files, box 1, RG 54.

14. Townsend, "Conditions Affecting," 173.

15. Hays, "Progress in Plant and Animal Breeding"; "Breaking and Establishing High Grade Pedigree Strains of Sugar Beet Seed."

16. "Sugar: History of Appropriations, 1883 to 1946," SPI, Records Relating to the History of Sugar Plant Investigations, box 1, RG 54.

17. Townsend, "Synopsis of Projects, 1910," SPI, Townsend Office Files, box 3, RG 54.

18. Wyckoff, *Creating Colorado*, 101–53.

19. The results of CAC research on Holly's beets were later published in Headden, *Fixation of Nitrogen*; and Headden, *Deterioration in the Quality*.

20. Karl Kellerman to W. A. Orton, March 16, 1910; and Orton to Spillman, April 30, 1910, SPI, Townsend Office Files, box 4, RG 54.

21. Scott, *Reluctant Farmer*, 254–87.

22. It is easy to overstate this disagreement. Headden had been working on the problem of excess nitrates in sugar beet fields for over a decade. While he concluded that excess nitrates—naturally occurring or applied as fertilizer—diminished sugar content in beets, he still concluded that nitrogen from barnyard manure was effective in increasing sugar beet yields. Spillman, for his part, offered speculative conclusions about the soils of the Southern Colorado Piedmont based on historical observations of nitrogen depletion elsewhere. In fact, SPI, Headden, and other CAC experiment-station researchers worked collaboratively in the beet fields of Southern Colorado throughout the course of investigations.

23. Spillman to Warren, April 4, 1910, SPI, Townsend Office Files, box 4, RG 54.

24. Spillman to Warren, April 4, 1910, and Spillman to Warren, n.d., SPI, Townsend Office Files, box 4, RG 54.

25. Harold Powell to Louis Carpenter April 30, 1910, SPI, Townsend Office Files, box 4, RG 54.

26. Spillman to William Wiley, April 4, 1910, SPI, Townsend Office Files, box 4, RG 54.

27. In 1910 Townsend possessed a more comprehensive knowledge of all aspects of sugar beet production than any other American. He published on topics

such as beet crop rotations, production costs, beet seed industry history, fertilizer in beet production, yields and sugar content, and mapping out regions where beet sugar agriculture was possible.

28. Warren to Wiley, April 24, 1910, SPI, Townsend Office Files, box 4, RG 54.

29. "Memo Regarding the Sugar Beet Industry in the United States," Jan 11, 1916, SPI, Townsend Office Files, box 3, RG 54.

30. United States Tariff Commission, *Sugar*, 19–39.

31. The relationship between war, calories, and energy is detailed in Cullather, "Foreign Policy of the Calorie," 337–64.

32. Pearl, "Memorandum for Sugar Equalization Board," 230–32. Some of this angst over sugar rationing was likely the result of skyrocketing consumption prior to the war. According to evidence cited by Sidney Mintz, in the sixty years leading to World War I in the United States: "average daily per-capita consumption of sugars as a proportion of carbohydrates . . . increased from 31.5% to 52.6%." See Mintz, *Sweetness and Power*, 199.

33. Bernhardt, *Government Control of Sugar*, 25–67.

34. Ellis et al., *Tariff on Beet Sugar*, 23–43.

35. United States Tariff Commission, *Sugar*, 216–18.

36. On the push for efficiency in American agriculture see Fitzgerald, *Every Farm a Factory*. On how World War I jumpstarted the chemical industry in the United States and forged ties with agriculture see Russell, *War and Nature*.

37. Examples of public scrutiny of labor practices include Clopper and Hine, *Child Labor in the Sugar-Beet Fields*; Coen, *Children Working on Farms*; Matthews, Light, and United States, *Child Labor and the Work of Mothers*; and Johnson, *Welfare of Families*.

38. Those factories were located in Brighton, Ovid, and Johnstown.

39. On growth of the chemical industry, especially the effect that World War I had on the agrochemicals, see Russell, *War and Nature*, 37–94; Chandler, *Shaping the Industrial Century*, 41–82.

40. Robert Harveson, "Cercospora Leaf Spot of Sugar Beets," accessed November 2, 2021, http://ianrpubs.unl.edu/live/g1753/build/g1753.pdf.

41. Anton Skuderna to W. W. Tracey, November 13, 1925, NAL, box 2.

42. The Hagley library contains numerous advertisements and publications for DuPont's chemical products. Most are individually catalogued and shelved. I consulted multiple publications for each product referred to here. For footnote brevity, I will cite a representative number. On Semesan, see "DuPont Semesan: The Premier Seed Disinfectant for All Agricultural Purposes," 1926, and "Dupont Information Bulletin, 25 August 1927," Hagley Library;

"Treated Seed Experiment, M.L. Reeder Farm—Brigham City, Utah," April 23, 1925, NAL.

43. Works that show the collaboration between state-sponsored science and the chemical industry in the post–World War II era include Dunlap, *DDT*, and Davis, *Banned*.

44. Examples include Hundley, *Water and the West*; Worster, *Rivers of Empire*; Pisani, *Water and American Government*.

45. On the extent to which irrigation had been developed on the Piedmont by the early twentieth century, see Mead, *Irrigation Institutions*, 143–79. For a study on irrigation along the Cache la Poudre River see Laflin, *Irrigation, Settlement, and Change*.

46. Hemphill and United States, *Irrigation in Northern Colorado*; "Colorado Uses Less than Half its Available Irrigation Water," Third Conference on Irrigation, Fort Collins CO, 1930, Irrigation Research Papers, box 5; A. H. Outler, "Jackson Lake," unpublished typescript, n.d., Fort Morgan Museum.

47. Kezer, *Irrigated Farming in Colorado*; Hemphill and United States, *Irrigation in Northern Colorado*.

48. G. H. Palmes, "Ralph Parshall—The Man" (speech, October 12, 1949), "Professional Bibliography: R.L. Parshall," August 6, 1958, Ralph Parshall Papers, box 1; "Memorandum Concerning the Request for a Grant of Public Works Administration Finds for the Construction of an Addition to the Irrigation Hydraulic Laboratory of the Colorado Agricultural College at Fort Collins," May 21, 1935, Irrigation Papers, box 16, Addition to the Laboratory.

49. In irrigated agriculture, seepage refers to water moving through porous soils rather than being taken up by plant roots. Seeping water often returns to rivers, creating what farmers and irrigation engineers call return flows. Return flows could be used again downstream. For Parshall, seepage indicated poor use of water, and return flows that were not used by other farmers represented waste in the system.

50. Parshall, *Return of Seepage Water*; Parshall, "The Improved Venturi Flume," March 18, 1927, Irrigation Research Papers, box 7, KOA Radio Talks, 1931–1935.

51. Parshall to Gillette, April 30, 1925, Irrigation Research Papers, box 5.

52. Mead, *Irrigation Institutions*, 143–79.

53. Parshall, "The Improved Venturi Flume"; Parshall, "Importance of Measuring Irrigation Water," Parshall Papers, box 1, KOA Radio Talks, 1931–1935.

54. Parshall, "The Improved Venturi Flume"; Chris Thornton, Director of the Engineering Research Center and Hydraulics Laboratory at Colorado State University. Interview by author, December 22, 2015, Fort Collins CO; Parshall, untitled, March 14, 1932, Parshall Papers, box 1, KOA Radio Talks, 1931–1935.

55. Parshall, "The Improved Venturi Flume"; Thornton interview; Parshall, untitled, March 14, 1932.

56. Thornton interview.

57. Parshall, "Suggestions for Increasing our Irrigation Supply," September 14, 1931; March 14, 1932; April 1, 1932; May 1, 1933, Parshall Papers, box 1, KOA Radio Talks, 1931–1935.

58. Parshall, *Measuring Water in Irrigation Channels*; Parshall, *Parshall Flumes of Large Size*.

59. For an example, see Keirnes, *Water*, 6–7; Thornton interview.

60. Thomas Trout, retired supervisory engineer, USDA, interview by author, Fort Collins CO, July 25, 2017.

5. MECHANIZATION AND ENGINEERING

1. Thomas F. Mahony Papers, Correspondence and Clippings, 1909–1928, TFM, box 1, folders 1–5; Donato, *Mexicans and Hispanos*, 14–16.

2. Mahony, "Wages of the Unskilled Workers in Colorado," May 27, 1929, TFM, box 2, folder 17.

3. *New Republic*, August 14, 1929, 323–24; "Pauper Labor and Child Labor," *The Nation*, July 8, 1931. The work of Mahony and the Mexican Welfare Committee also contributed to various government publications, academic works, and the writings of reformers such as Carey McWilliams. See for example, "The Protection of the Children of Seasonal Migrant Workers," *Report Submitted by the International Labour Office of the Child Welfare Committee to the League of Nations*, April 4, 1932, TFM, box 3, folder 2; "Colorado White House Conference on Children in a Democracy," March 1942, TFM, box 3, folder 2; Brown, Sargent, and Armentrout, *Children Working in the Sugar Beet*; Coen, *Children Working on Farms*; Carey McWilliams, *Ill Fares the Land*, 122–31; Taylor, *Mexican Labor in the United States*.

4. Mahony, *Report of Addresses Given at the First City Conference on Denver's Social Problems, June 6–8, 1928 Under Auspices of General Council of The Denver Community Chest, Denver, Colorado*, June 6–8, 1928, TFM, box 2, folder 2; C. V. Maddux to Thomas Mahony, October 9, 1929, TFM, box 1, folder 7.

5. Ballinger, *History of Sugar Marketing*, 23–32.

6. Ballinger, *History of Sugar Marketing*, 23–32.

7. Katherine Lenroot, "Statement of Economic Conditions and Child Labor in Families Employed in the Beet Fields of Colorado," August 11, 1933, TFM, box 2, folder 2.

8. "Labor Conditions Are Assailed in Northern Colorado," *Denver Post*, March 5, 1931; "Slavery Conditions in Beet Fields Reported to State," *Colorado Labor Advocate*, May 25, 1933.

9. Mahony to Senator Edward Costigan, February 22, 1931, TFM, box 1, folder 10.

10. "Labor Conditions are Assailed in Northern Colorado," *Denver Post*, September 5, 1931; "Slavery Conditions in Beet Fields Reported to State"; United States Congress, Senate, *Smoot Hawley Tariff Debates*; Mahony to Edward Costigan, February 22, 1931, TFM, box 1, folder 10.

11. For example, in 1924, Weld County doled out over $5,000. In 1930, Colorado Catholic Charities spent over $600,000 in aid for Mexicans. See Maddux to Mahony, March 11, 1925, TFM, box 1, folder 5; Mahony to Maddux, March 29, 1931, TFM, box 1, folder 10.

12. Lenroot, "Statement of Economic Conditions."

13. The size of Manual's contract implies a great deal about child labor. The average adult beetworker could handle ten acres in a season. So, his 1933 contract presumes that his wife and children would spend considerable time in the beet fields. See Lenroot, "Statement of Economic Conditions."

14. On the efforts of Piedmont beet laborers to organize unions beginning in the 1920s, see Vargas, *Labor Rights*, 27–34, 70–76, and 148–50.

15. Deutsch, *No Separate Refuge*, 162–99; Vargas, *Labor Rights*, 70–76; Valdés, "Settlers, Sojourners, and Proletarians," 110–23.

16. Mae Ngai has argued that immigration law during the period criminalized Mexicans, who were viewed by society as unfit and morally outside of societal norms. This made their presence in the United States subject to whims and individual judgments, such as occurred on the Piedmont during the 1930s. See Ngai, *Impossible Subjects*, 56–90.

17. On the work of the ILD in the United States during the 1930s, see Goodman, *Stories of Scottsboro*. For an excellent example of communist-led labor organizing among minority groups during the same period, see Kelley, *Hammer and Hoe*.

18. Valdés, "Settlers, Sojourners, and Proletarians."

19. C. V. Maddux to Mahony, April 15, 1930, and Mahony to Maddux, April 20, 1930, TFM, box 1, folder 10; Mahony to Cleofas Calleros, March 3, 1928, TFM, box 1, folder 5; Mahony to Calleros, April 6, 1935, TFM, box 1, folder 16.

20. Those basic commodities were corn, wheat, milk, cotton, rice, peanuts, and tobacco. For a more thorough discussion of the provisions of the AAA, see Conkin, *A Revolution down on the Farm*, 59–68.

21. Ballinger, *History of Sugar Marketing*, 23–33; Merleaux, *Sugar and Civilization*, 174–86.

22. Leo Rodriguez, "To the Great Western Sugar Company," *Loveland Press*, May 25, 1926.

23. Lenroot, "Statement of Economic Conditions"; "Testimony of Leo Rodriguez on Behalf of Those Employed in Growing Sugar Beets," Agricultural Adjustment Act Hearings, reel 10, docket no. 22, August 11, 1933, 441–44.

24. Bressette to Mahony, August 12, 1933, TFM, box 1, folder 14; "Colorado Beet Workers Committee to Henry A. Wallace," March 20, 1934, TFM, box 1, folder 15. For a summary of Rodriguez's testimony and the efforts of the National Child Labor Committee and the Catholic Conference on Industrial Problems on behalf of beetworkers in 1933, see Merleaux, *Sugar and Civilization*, 176–79.

25. Heston, *Sweet Subsidy*, 75–116; U.S. Department of Agriculture, Office of Information, Press Service, "Sugar Contracts to Protect Labor; Growers' Benefits to Total $10,000,000," July 3, 1934 (Washington: Agricultural Adjustment Administration), TFM, box 3, folder 3; U.S. Department of Agriculture, Office of Information, Press Service, "Minimum Wages Set for Sugar Beet Field Labor in Western Areas," April 20, 1935 (Washington: Agricultural Adjustment Administration), TFM, box 3, folder 3.

26. Heston, *Sweet Subsidy*, 75–116; U.S. Department of Agriculture, Office of Information, Press Service, "Sugar Contracts to Protect Labor"; U.S. Department of Agriculture, Office of Information, Press Service, "Minimum Wages Set"; Colorado Beet Workers Committee to Wallace, August 22, 1934, TFM, box 1, folder 15.

27. E. L. Kirkpatrick, "Statement Presented at Hearing on Proposed Marketing Agreement for the Stabilization of Sugar," August 10, 1933 (Washington: Federal Emergency Relief Administration), TFM, box 1, folder 14; Mahony to Paul Taylor, July 14, 1934, TFM, box 1, folder 15; Harry Hopkins to Edwin Johnson, November 5, 1934, TFM, box 1, folder 15; "Slash in State WPA Workers is Underway," *Rocky Mountain News*, March 13, 1936.

28. Those labor shortages existed primarily because Mexicans had either migrated to the Southwest or Mexico or were forcibly repatriated to Mexico, in one the largest mass deportations in American history. On that repatriation, see Balderrama and Rodriguez, *Decade of Betrayal*.

29. "Protesting Intimidation of Beet Workers on Relief," January 1937, *Rural Worker*, TFM, box 2, folder 20.

30. It bears pointing out that many examples of exploitation remained. See Larson, *Beet Workers on Relief*.

31. Heston, *Sweet Subsidy*, 75–116; Ballinger, *History of Sugar Marketing*, 32–39.

32. May, *Great Western Sugarlands*, 187–89; Macy et al., *Changes in Technology*; Rasmussen, "Technological Change," 31–36; Moorhouse et al., *Farm Practice*; *Colorado: Report on the Farmers' Costs*. According to the U.S. Tariff Commission, out of 583 farms investigated in Colorado in 1922, with the majority on the Piedmont, less than 10 percent (51) harvested more than 50 acres and only 17 (less than 3 percent) of farms harvested more than 81 acres, which was the average across the sugar beet industry in California.

33. Macy et al., *Changes in Technology*, 17–18; GWS began to encourage the application of phosphates in the late 1920s, see also H. W. Dahlberg, "Fertility Requirements of Beets and Relation between Manure and Superphosphates," *TTL*, 1930, 85–86.

34. Macy et al., *Changes in Technology*; Coons, "Sugar Beet," 190.

35. "Dr. Brewbaker Joins Company," *TTL*, 1937, 86.

36. Brewbaker, "Why Thin Sugar Beets," *TTL*, 1937, 84.

37. "Dr. Brewbaker Joins Company."

38. Campbell and Cattanach, "American Society of Sugar Beet Technologists."

39. "Sugar Beets Mechanized," October 1946, SPI, Farm Power and Machinery, box 1, RG 54.

40. D. J. Roach, "You Can Cut Thinning Labor in Half!," *TTL*, 1943, 3; Roy Bainer, "Use of Sheared Seed on the Increase," *TTL*, 1943, 6.

41. Some of those laborers were replaced by Mexican braceros, who were especially prominent during thinning season. While their numbers were greatest during the last two years of World War II, Piedmont growers continued to employ braceros until 1964, when the program ended and the industry was almost entirely mechanized. On Piedmont braceros, see Crisler, "Mexican Bracero Program."

42. "Sugar Beets Mechanized."

43. United States Department of Agriculture, Office of Administrator of Research and Marketing Act, and Sugar Advisory Committee, *Sugar Problems*, 13.

44. Brewbaker, "Single Germ Beet Seed a Commercial Possibility," *TTL*, 1950, 29–31; United States Department of Agriculture, Office of Administrator of Research and Marketing Act, and Sugar Advisory Committee, *Sugar Problems*; "Recent Developments in Sugar Beet Breeding," March 21, 1951, NAL, box 1; F. V. Owen, "Opportunities for Improving Sugar Beets," December 4, 1957, NAL, box 8, unfiled.

45. Compounds in use included 2,4-Dichlorophenoxyacetic acid (2,4-D), Dichlorodiphenyltrichloroethane (DDT), and Isopropyl-N-Phenyl-Carbonate (IPC). For a fuller discussion of chemicals used on the Piedmont, see chapter 6. On

field trials, see for example G. W. Deming, *Progress Report on Weed Control Studies at Ft. Collins, Colo.*, 1948, NAL, box 8, unfiled, and Albert Murphy, "Effect of DDT on Beet Leafhoppers, Curly Top, and Yields of Sugar Beet Varieties," NAL, box 10.

46. Moorhouse and Summers, *Saving Man Labor in Sugar-Beet Fields*; Rasmussen, "Technological Change"; May, *Great Western Sugarlands*, 198–99; *Report of the 1951 Evaluation Tests*, n.d., NAL, box 5.

47. Unlike the Southern Colorado Piedmont, the Northern Piedmont was not among the regions whose soils were decimated by the Dust Bowl. For a recent discussion on the Dust Bowl there, see Sheflin, *Legacies of Dust*.

48. Originally, it was called the Grand Lake Project. For narrative uniformity, I will use the name given to it in 1937, the Colorado-Big Thompson Project. For an exhaustive history of the project, see Tyler, *Last Water Hole*.

49. To grow an acre of beets in Colorado in the mid-1920s, growers desired between 2 and 2.5 acre-feet. Experiments in the early 1930s suggested that yields would increase with slightly more water. Thus, I calculated water requirements for an acre of sugar beets at 2.5 acre-feet. During 1936, farmers on the Piedmont contracted to grow 162,000 acres. See *TTL*, January 1936, 3; Fortier, "Irrigation Requirements," 29.

50. Pfaff, *Colorado-Big Thompson*, 76–87; Tyler, *Last Water Hole*, 3. It is also notable that the Colorado River Compact, which allocated 7.5 million acre-feet of water from the Colorado River to be shared between Colorado, Wyoming, Utah, and New Mexico, had been ratified by Congress in 1922. While the Compact legitimated Colorado's claim to its namesake river, it was rarely mentioned in C-BT discussions during the 1930s. On the Compact, see Hundley, *Water and the West*.

51. Historians Donald Pisani argues that Reclamation had primarily become a construction agency by the 1930s, distancing itself from founding goals. His peer, Donald Worster, on the other hand, argues that supporting small farmers and opening new lands to agricultural development had never animated Reclamation. Rather, he claims that exploiting nature for private profit was the agency's aim from the start. See Pisani, *Water and American Government*; Worster, *Rivers of Empire*.

52. Tyler, *Last Water Hole*, 26–36; Pfaff, *Colorado-Big Thompson Project*, 76–87.

53. Parshall, *Agricultural Economic Summary*.

54. N. R. McCreery, "What More Water Means to You," *TTL*, July 1936, 110; Hemphill and United States, *Irrigation in Northern Colorado*, 1922.

55. Parshall, *Agricultural Economic Summary*.

56. Parshall, *Agricultural Economic Summary*.

57. For one example of Parshall's misgivings about trans-mountain diversion, see Parshall to M. R. Lewis, September 21, 1930, Irrigation Research Papers, box 5.

58. In 1938, when the C-BT passed Congress, water users on the Piedmont paid one dollar and fifty cents per acre-foot of water on demand. Parshall points out that from 1925–34 in the Cache la Poudre Valley, the average cost of water on demand was four dollars per acre-foot. See Tyler, *Last Water Hole*, 102; Parshall, *Agricultural Economic Summary*, 32–33.

59. Keirnes, *Water*, 42.

60. Brewbaker, "Studies of Irrigation Methods for Sugar Beets in Northern Colorado," and Brewbaker, "Better Stands of Sugar Beets," *TTL*, 1934, 2.

61. Examples of promotion by Great Western Sugar include McCreery, "What More Water Means to You," *TTL*, July 1936; "Sugar Beets Tell the World," *TTL*, January 1937; "Transmountain Water Diversion Worth Millions to Growers," *TTL*, March 1937; "Transmountain Water Will Help Pumping Plants," *TTL*, June 1937.

62. Autobee, *Colorado-Big Thompson Project*, 17. For a list of C-BT features, including construction dates, see Pfaff, *Colorado-Big Thompson Project*, 76–87.

63. The C-BT also generated power at several of its dams, most of which was distributed to Piedmont cities and to Rocky Mountain National Park. See Tyler, *Last Water Hole*, 58–127.

64. Tyler, *Last Water Hole*, 58–98.

6. BUILDING THE PETROCHEMICAL PARADIGM

1. Tyler, *WD Farr*, 28–52.

2. The "Big Five" meatpackers consisted of Swift, Armour, Cudahy, Morris, and National. On their development, see Specht, *Red Meat Republic*, 174–214; Skaggs, *Prime Cut*, 90–129.

3. For some history on the Monfort family and W. D. Farr, see Barnhart, *Kenny's Shoes*; Tyler, *WD Farr*, 3–76; W. D. Farr, interview by Peggy Ford, November 29, 2004, transcript, City of Greeley Museums, Greeley CO. On cattle feeding in Northern Colorado during the early twentieth century, see Bussing, "Cattle Feeding Industry," 17–73.

4. On how Estes Park was constructed to accommodate the tourist trade in Rocky Mountain National Park, see Frank, *Making Rocky Mountain National Park*.

5. On the relationship between Farr and Monfort, including their fence-rail conversation, see W. D. Farr, interview by Peggy Ford.

6. In chapter 2, I describe the Piedmont agroecological model and its relationship with capitalism in detail. For another example of the historical interde-

pendence between cattle and crops, see Duncan, *Centrality of Agriculture*, 63–71.

7. Historians Adam Romero, Steven Stoll, and Frederick Davis point out that farmers in places such as California and Washington were using a variety of synthetic insecticides and fertilizers prior to 1940. Romero shows that these fractured, though did not break, existing crop rotations. By contrast, Great Western often told its growers that the best way to avoid insect infestation was to continue existing crop rotations. See Stoll, *Fruits of Natural Advantage*, 94–123; Davis, *Banned*, 1–37; Romero, "'From Oil Well to Farm,'" 70–93.

8. Phosphorous provides a mild exception to the rule, as Piedmont farmers began to employ commercial phosphorous fertilizer in small quantities during the late 1920s. For an example of Great Western Sugar's phosphate experiments and promotion, see Henry Dahlberg, "Four Western District States to Profit from Phosphate Fertilizer Experiments," *TTL*, April 1929. On the growth and importation of phosphates, see Cushman, *Guano*, 10–22. Potassium was rarely added to Piedmont soils. As historian Paul Conkin explains, clay soils, such as were common on the Piedmont, are storehouses of potassium. See Conkin, *Revolution down on the Farm*, 108–19.

9. On the work of industry and agricultural colleges in the development of seed disinfectants and machinery designed for the sugar beet industry, see chapters 4 and 5. Colorado Agricultural College (CAC) became Colorado A&M in 1935, and then was renamed Colorado State University in 1957. For narrative uniformity, CAC will be used throughout this chapter. On the growth of petrochemical companies in the twentieth century, including Dow, DuPont, and Shell, see Chandler, *Shaping the Industrial Century*, 41–143.

10. European chemical companies developed organic chemicals using coal tars starting in the 1880s, while the transition toward organic chemicals among American companies did not occur until the 1910s. See James E. McWilliams, *American Pests*, 111–67; Mart, *Pesticides, A Love Story*, 11–30; Commoner, *Closing Circle*, 125–39; Davis, *Banned*, 39–71; Chandler, *Shaping the Industrial Century*, 41–82; Romero, "'From Oil Well to Farm.'"

11. On the Haber-Bosch Process, see Smil, *Enriching the Earth*; Leigh, *World's Greatest Fix*, 139–54.

12. Commoner, *Closing Circle*, 132.

13. Russell, "The Strange Career of DDT," 770–96; Dunlap, *DDT*, 59–75; Rudd, *Pesticides and the Living Landscape*, 61–62; Perkins, *Insects, Experts, and the Insecticide Crisis*, 3–23; Davis, *Banned*, 39–71.

14. W. D. Farr, speech at the Billings Farmers Institute, February 14, 1952, W. D. Farr Collection, box 1.

15. Gordon Mickle, *Annual Extension Entomologist Report for 1946*, A E X T, box 116; Dove, "Livestock Insect Control."

16. Mickle, *Annual Extension Entomologist Report*; "Flies Make Profits Fly," *Down to Earth*.

17. Albert Goodman, *Veterinarian Specialist Extension Report for 1947*, A E X T, box 127.

18. Russell, "Strange Career of D D T"; Rudd, *Pesticides and the Living Landscape*, 61–62, 103–10, 154–71.

19. Rudd, *Pesticides and the Living Landscape* 141–48.

20. Cuff, "Livestock Pest Control." On the use of Dow products on cattle grubs, as a replacement for D D T, see *Over the Feed Bunk*, August 1956, Records of the Colorado Cattle Feeders Association.

21. Lindquist, "Livestock Insect Control"; Knipling, "Arthur W. Lindquist," 135–36.

22. Robert Van Den Bosch was one of the few examples of agri-state scientists working on biological control. During a career at U C Berkeley, which began in 1946, he focused on "integrated pest management" that used a combination of farm monitoring, employing a combination of natural pest predators, carefully selecting planting dates, using row crops that functioned as insect breaks, and limited pesticides applications. According to Van Den Bosch, these methods were quite effective even though he was criticized for not adopting the "pesticide paradigm." See Van Den Bosch, *Pesticide Conspiracy*.

23. By the time Lindquist assumed a prominent role in B E P Q, economic entomologists dominated the field of entomology. These scientists regarded the majority of insects as enemies of agriculture. Their research therefore focused on killing insects. Consequently, they rarely questioned the value of chemicals. See Dunlap, *D D T*, 17–38.

24. The majority of Piedmont feeders fed less than five hundred cattle annually, well under the one thousand head needed to be classified as a commercial feeder. This suggests that the majority had not abandoned crop farming entirely. Recognizing the importance of diversified farming for sugar beet production, Great Western Sugar sought to convince farmer/feeders that they would be more successful if they continued to rotate crops and raise cattle. See *Over the Feed Bunk*, January and March 1957, Records of the Colorado Cattle Feeders Association.

25. Don Ament, cattle feeder and former Colorado Agricultural Commissioner, interview by author, June 23, 2019, Iliff C O.

26. Richard Seaworth, owner, Seaworth Farms, interview by author, June 21, 2019, Wellington C O; Richard Maxfield, former cattle buyer and livestock

commission man for the Denver Union Stockyards, interview by author, June 25, 2019, Greeley CO; Chuck Sylvester, former director of the National Western Stock Show, interview by author, June 20, 2019, La Salle CO; Kent Peppler, past president of the Rocky Mountain Farmers' Union, interview by author, June 18, 2019, Mead CO.

27. Proceedings of the Agricultural Chemicals Dealers Conference, Panel on Insecticides; To analyze how the beet sugar industry researched and talked about DDT, I examined the proceedings of the 1946, 1948, and 1950 American Society of Sugar Beet Technologists (ASSBT), available from https://www .bsdf-assbt.org/proceeding; J. R. Douglass and Albert Murphy, "Effect of DDT on Beet Leafhoppers, Curly Top, and Yields of Sugar Beet Varieties," Proceedings of the Seventh Biennial Meeting of ASSBT (Salt Lake City UT, February 7, 1952).

28. During the first half of the twentieth century, the field of entomology was dominated by economic entomologists who focused on insect control and elimination. On how economic entomology came to dominate the field and its consequences, see Dunlap, DDT, 17–38; Davis, Banned, 39–71; McWilliams, American Pests, 111–43.

29. In fact, according to Dale Wolf, who received his PhD in weed science in 1948 from Rutgers University, he may have been the first person in the United States to receive such a degree. Wolf spent the majority of his career working for DuPont's agricultural chemical division. See Dale Wolf, interview by Hounshell and Smith October 8, 1986, Hounshell and Smith Oral Histories, Acc. 1878, box 2.

30. "The Herbicidal Action of 2,4 Dichlorophenoxyacetic and 2,4,5 Trichlorophenoxyacetic Acid on Bindweeds," 154–55; O. A. Holkesvig, "Effects of 2,4-D on Sugar Beets," Proceedings of the Sixth Biennial Meeting of ASSBT (Detroit MI, March 7, 1950).

31. G. W. Deming, "Use of IPC for Weed Control in Sugar Beets," Proceedings of the Seventh Biennial Meeting of ASSBT (1952); R. T. Nelson, "Grass Control in Sugar Beets with the Herbicides IPC, TCA, and DCU," Proceedings of the Eighth Biennial Meeting of ASSBT (Denver, February 3, 1954); Roger Blough, Jess Fultz, and John O. Gaskill, "Control of Annual Grass Weeds in Sugar Beets by Sodium 2,2-dichloropropionate," January 24, 1956, NAL, box 1.

32. Fred Fletcher, untitled lecture given at the Rocky Mountain Conference of Entomologists, 1940, Records of the Rocky Mountain Conference of Entomologists; George L. James, Weld County Agricultural Extension Specialist Report, 1950, AEXT, box 92; Romero, "'From Oil Well to Farm'"; Chandler, Shaping the Industrial Century, 41–82.

33. The various forms of pelleted and liquid nitrogen available to farms were byproducts of wartime explosives and the Depression-era Tennessee Valley Authority. William R. Farr recalls inserting anhydrous ammonia into his family's Piedmont irrigation ditches by the 1950s. See Conkin, *Revolution down on the Farm*, 110–11; William R. Farr, Farr Farms and Feedlots, interview by author, July 5, 2019, Greeley CO; Agricultural Chemicals Dealers Conference, Panel on Insecticides; Gardner and Robertson, *Nitrogen Requirement of Sugar Beets*; Davan, Schmehl, and Stewart, *Fertilizer Use and Trends*.
34. "Relation of Commercial Fertilizer to Yield of Beets, 1946–1953," Records of the Great Western Sugar Company, accession 2, series 5, box 2.
35. Floyd, Schmehl, and Stewart, *Fertilizer Use and Trends*.
36. James, *Weld County Extension Report*.
37. *1957 Insect Control Recommendations*; Thornton and Colorado Agricultural Experiment Station, *Report on Herbicides*.
38. Peppler, interview; additional observations are taken from interviews by the author in June 2019.
39. On the mission of land grant colleges, see Rosenberg, *No Other Gods*, 153–72. For further examples of how research vigor was rapidly shifting from land grant colleges to industry in the postwar period, see Freeborn, "How Experiment Stations Help Develop," 1; Dale Wolf interview.
40. J. L. Anderson, *Industrializing the Corn Belt*; Vail, *Chemical Lands*. Jim Hightower and Wendell Berry are two among several critics who argue that land grants failed in their mission to support farmers. Both argue that land grants became pawns of agro-industry and wealthy farmers. Berry, *Unsettling of America*; Hightower and Agribusiness Accountability Project, *Hard Tomatoes, Hard Times*.
41. Examples of this narrative include James, *Weld County Agricultural Extension Report*; U.S. Department of Agriculture and Production and Marketing Administration, *Pesticide Situation for 1952–53*.
42. Freeborn, "How Experiment Stations Help," 1; A summary of the comments on pesticides during the Delaney hearings is provided in Davis, *Banned*, 116–52.
43. Pollan, *Omnivore's Dilemma*, 15–64; Fitzgerald, *Business of Breeding*, 9–42; Mangelsdorf, "Hybrid Corn," 39–47; John Fraser Hart, *Changing Scale of American Agriculture*, 41–61.
44. Average first frost dates on the Piedmont varied, depending on location. For example, in Fort Collins it was September 18 and in Fort Morgan it was September 25. See Leonard, Fauber, and Tucker, *Tests with Hybrid Corn*.

45. McClymonds, *Colorado Corn Problems*. For an example of an early article that established the importance of sugar beets in feeding cattle and the dangers of importing corn, see Carlyle, *Feeding Steers*.

46. Fitzgerald, *Business of Breeding*, 9–132.; Brandon, Curtis, and Leonard, *Corn Production in Colorado*.

47. Mangelsdorf, "Hybrid Corn."

48. Fitzgerald, *Business of Breeding*, 9–132; Brandon, Curtis, and Leonard, *Corn Production in Colorado*.

49. Leonard, Fauber, and Tucker, *Tests with Hybrid Corn*; French and Fauber, *Tests of Hybrid Corn under Irrigation in Colorado in 1945*; Gausman and Fauber, *Tests of Hybrid Corn under Irrigation in Colorado in 1948*.

50. According to W. D. Farr's son William R. Farr, by the mid-1950s, Piedmont farmers were producing thirty to forty more bushels per acre than their corn farming peers in the Midwest. The gain, however, was offset by the costs associated with irrigation, generally absent from Midwestern farms. See William R. Farr, interview by author.

51. Corn grain consists of the kernels on the corn cob. Corn silage is the rest of the plant, generally ground up and fed to livestock.

52. Leonard, Fauber, and Tucker, *Tests with Hybrid Corn*; French and Fauber, *Tests of Hybrid Corn under Irrigation in Colorado in 1945*; Gausman and Fauber, *Tests of Hybrid Corn under Irrigation in Colorado in 1948*; Crumpacker, *Performance Tests of Hybrid Corn Varieties*; *Over the Feed Bunk*, February and September 1956.

53. Statistics were compiled from the Colorado Agricultural Census. The years 1939, 1944, 1949, 1954, and 1959 were used. See U.S. Department of Commerce, *Agriculture Census*; Leonard, Fauber, and Tucker, *Tests with Hybrid Corn*; French and Fauber, *Tests of Hybrid Corn under Irrigation in Colorado in 1945*; Gausman and Fauber, *Tests of Hybrid Corn under Irrigation in Colorado 1948*; Crumpacker, *Performance Tests of Hybrid Corn Varieties*; *Over the Feed Bunk*, February and September 1956.

54. W. D. Farr, interview by Peggy Ford; Maxfield interview.

55. Davan, Schmehl, and Stewart, *Fertilizer Use and Trends*.

56. Rod Ulrich, cattle feeder, interview by author, LaSalle co, June 27, 2019; William R. Farr, interview by author; Jenny, "Making and Unmaking of a Fertile Soil."

57. While there were certainly examples of insect infestation, such as the non-native European corn borer, it is generally true that insects were a much smaller concern than weeds. See Olmstead, *Creating Abundance*, 64–97.

58. Ament interview; Seaworth interview; "The Herbicidal Action of 2,4 Dichlorophenoxyacetic"; *1957 Insect Control Recommendations.*
59. Tyler, *Last Water Hole,* 216–60.
60. Jenny, "Making and Unmaking of a Fertile Soil"; Ament Interview; William R. Farr, interview by author; Peppler interview.

7. MANUFACTURING BEEF

1. *Research on Bovine Respiratory Disease: A Preliminary Report to the Research Committee of the Colorado Cattlemen's Association,* n.d., CCFA, box 6, folder 2 of 2.
2. Colorado Agricultural College became Colorado A&M in 1935 and in 1957 was renamed Colorado State University. For narrative uniformity, CSU will be used throughout this chapter.
3. William R. Farr, former manager, Farr Farms and Feedlots, interview by author, July 5, 2019, Greeley CO.
4. Progress and updates on Red Nose research were often provided in the Colorado Cattle Feeders Association bulletin, *Over the Feed Bunk.* See *Over the Feed Bunk,* November 1955, July 1956, August 1957, and September 1957; see also Matsushima, *Journey Back.*
5. In fact, by adopting the name Colorado State University, the school self-consciously sought to claim its place as a research-focused institution.
6. Kent Peppler, past president of the Rocky Mountain Farmers' Union, interview by author, June 18, 2019, Mead CO.
7. Feed-conversion ratio refers to the relationship between feed quantity and weight gain. Typically, it was expressed as the amount of feed required for one pound of gain. By 1970 the feedlot industry standard was less than eight pounds of feed for a single pound of gain.
8. Don Ament, cattle feeder and former Colorado Agricultural Commissioner, interview by author, June 23, 2019, Iliff CO.
9. By the 1950s most cattle arrived in feedlots via commercial trucks, which could meet the flexible demands of the industry more effectively than trains. See Hamilton, *Trucking Country,* 136–62.
10. On the time cattle spent in feedlots and the processes that occurred there, see W. D. Farr, "Ceneca International Symposium," Paris, March 1967, W. D. Farr Collection, box 1; "Production of Monfort," unpublished typescript, JBS, Monfort Collection, box 5, Scripts.
11. W. M. Beeson, "Excerpts from Remarks of Dr. W. M. Beeson, Professor of Animal Science, Purdue University," Meeting of the American National

Cattlemen's Association, Omaha NE, January 14, 1959, W. D. Farr Collection, box 18, Miscellaneous Good Speeches.

12. Initially, the most robust work on diseases and how to achieve rapid weight gain on less feed were conducted on chickens during the pre–World War II period. Much of the later work on swine and cattle capitalized on this work. See William Boyd, "Making Meat."

13. Boyd, "Making Meat"; Summons, "Animal Feed Additives," 305–13; W. M. Beeson, "Excerpts."

14. Summons, "Animal Feed Additives"; Marcus, *Cancer from Beef*; Langston, *Toxic Bodies*, 61–82; William R. Farr, interview by author; W. D. Farr, "Future Developments in Cattle Feeding," April 24, 1959, Indiana Cattle Feeders Day, W. D. Farr Collection, box 1. *Over the Feed Bunk*, January 1956 and February 1957; "W.D. Farr: Feedlot's 1969 Commercial Feeder of the Year," *Feedlot*, February 1969.

15. "Colorado Cattle Feeders Association Articles of Incorporation," CCFA, box 4, CCFA Permanent File; "Agreement between Colorado Cattle Feeders Association and Colorado Agricultural and Mechanical College," March 1955, copy obtained from Bill Hamerich, CEO, Colorado Livestock Association, June 20, 2019.

16. *Over the Feed Bunk*, August 1957; "Journal Drug Account, 12/31/57," CCFA, box 4.

17. For 1957, I correlated drug sales from pharmaceutical companies to CCFA with ads in *Over the Feed Bunk*. By the 1960s, CCFA was no longer a significant seller of feedlot drugs as those were either supplied through feed stores, veterinarians, or mixed and sold by large commercial feeders. So, I correlated industry membership in CCFA with ads in *Over the Feed Bunk*. See *Over the Feed Bunk*, 1957, 1958, and 1967; "Journal Drug Account, 12/31/57"; "Colorado Cattle Feeders Association Membership List, 1967–1968," CCFA, box 4.

18. CCFA, "Drugs, June 1967"; "Colorado Cattle Feeders Association Membership List"; "Journal Drug Account, 12/31/57."

19. "Jim Svedman's income and expenses for 1964, 1968, and 1970," Jim Svedman Papers, box 4, Farm, Ranch, Bookkeeping and Tax Record, Partnership and Bills and Inventories 1968 and 1970.

20. *Over the Feed Bunk*, December 1955 and May 1956. On the connection between DES and pregnancy in humans and cattle see Langston, *Toxic Bodies*.

21. Svedman Papers, box 1, Crops 1959–1975.

22. William R. Farr, interview by author; Madden and Hunter, *Economies of Size*.

23. W. D. Farr, "Cattle Feeding a Dynamic Industry," January 1969; speech at Michigan State University, February 2, 1966; and "Ceneca International Sym-

posium," Farr Cattle Talks, 1951–1986, and speeches, W. D. Farr Collection, box 1; Tyler, *WD Farr*, 74.

24. William R. Farr recalls a steady stream of visitors at Farr Farms during the 1950s and 1960s, including feeders, politicians, media, and others who were just curious. See William R. Farr, interview by author.

25. W. D. Farr, "Speech at Iowa Cattle Feeders Day"; W. D. Farr, "Future Developments in Cattle Feeding," April 24, 1959, Farr Cattle Talks, 1951–1986, W. D. Farr Collection, box 1.

26. Feed-conversion ratio refers to the quantity of feed required to add weight onto cattle. It was measured in pounds. By the 1960s, feeders sought a feed conversion rate of less than eight pounds of feed for a single pound of gain.

27. W. D. Farr, "Future Developments in Cattle Feeding"; W. D. Farr, "Economics of Cattle Feeding: Where We Have Been and Where We are Going," Washington State University, December 11, 1961, Farr Cattle Talks, 1951–1986, W. D. Farr Collection, box 1; W. D. Farr, "Cattle Feeding: A Dynamic Industry," January 1969, Farr Cattle Talks, 1951–1986, W. D. Farr Collection, box 1.

28. W. D. Farr, "Speech at Iowa Cattle Feeders Day."

29. John Matsushima, former animal sciences professor, interview by author, Colorado State University, June 28, 2018; John Matsushima, interview by James Hansen, March 2, 2010, Archives and Special Collections, Morgan Library, Colorado State University; Matsushima, *Broad Horizon*, 7–55.

30. Historian Alan Marcus argues that these networks were part of a larger attempt to increase the prestige and profile of land grant colleges while attracting an increasing number of university students, following World War II. While Jim Hightower acknowledges the growth of land grants in the 1950s and 1960s, he argues that in courting agro-industry, land grants abandoned a core mission to serve a broad swath of farmers to serve elite, corporate interests in rural America. See Marcus, *Service as Mandate*, 1–18; Hightower and Agribusiness Accountability Project, *Hard Tomatoes, Hard Times*, 1–6.

31. Matsushima, interview by author; Matsushima, interview by James Hansen; Matsushima, *Broad Horizon*, 56–58.

32. Ray Chamberlain, former president of Colorado State University, interview by author, July 11, 2017; Robert Zimdahl, professor emeritus of weed science at Colorado State University, interview by author, June 14, 2016, Fort Collins CO. For a broad indictment on the relationship between land grant colleges and industry, see Hightower and Agribusiness Accountability Project, *Hard Tomatoes, Hard Times*, 1–148.

33. Matsushima, *Broad Horizon*, 56–58.

34. Matsushima, interview by author; Matsushima, interview by James Hansen; Matsushima, *Broad Horizon*, 56–58.

35. Matsushima, interview by author; Matsushima, interview by James Hansen; "1975 National Western Fed Beef Contests," CCFA, box 2, National Western Fed Beef Contest; "Fat Cattle Can't Win: Beauty More than Skin Deep," *Denver Post*, January 15, 1974.

36. "Grass-fed Beef Taste Tests Show Less Acceptance," *AgriSearch: A Bi-Monthly Publication of the Colorado State University Experiment Station*, December 1975.

37. Matsushima, interview by author.

38. Chuck Sylvester, former Director of the National Western Stock Show, interview by author, June 20, 2019, La Salle CO.

39. Chuck Sylvester interview; Don Ament, June 23, 2019, interview by author; Hamerich interview.

40. Matsushima, *Broad Horizon*, 66–67, Matsushima, interview by author.

41. "Beef Nutrition Project," AEXT, box 80; D. D. Johnson to J. P Jordan, May 20, 1977, AEXT, box 80.

42. A commercial feedlot consisted of one thousand or more cattle.

43. Matsushima, interview by author.

44. "Metabolic Laboratory and Feedlot Nutrition Unit, Beef Cattle-Nutrition Research Activities, Animal Science Department," Records of the Colorado Cattlemen's Association, box 70.

45. EPA regulations published in 1972 stated that commercial operators, possessing more than one thousand cattle, could not discharge wastes into local waterways and could only do so after significant storm events, and only after applying for a permit to do so.

46. "Metabolic Laboratory and Feedlot Nutrition Unit." On August 4, 1972, DES was phased out of use in cattle by the FDA since it was a known carcinogen, thus violating the Delaney Clause of 1958. Federal courts upheld the ban in 1976. See Langston, *Toxic Bodies*, 83–133.

47. Matsushima, interview by author.

48. Madden and Hunter, *Economies of Size*; Bussing, "Cattle Feeding Industry," 6; Madsen, *Response of Colorado Cattle Feeders*.

49. Jack Guinn, "John Matsushima"; "History and Development of Farming Operations," Monfort Collection, box 6; "Production of Monfort."

50. Hamilton, *Trucking Country*, 144; "Production of Monfort"; "Monfort: The Legend," *Chilton's Commercial Car Journal*, September 1978, 101–10.

51. "Production of Monfort"; Mayer, "Monfort Is a One-Company Industry"; *Research on Bovine Respiratory Disease*.

52. Green chop alfalfa is alfalfa that has not been dried. Both green chop alfalfa and corn silage were cheap feeds that were much closer to the natural grasses that cattle were accustomed to and had evolved to eat. As such, they were used to transition feedlot cattle to corn grain.

53. Leonard Roe, "They Feed 200,000 Steers," *Monfort of Colorado*; "Production of Monfort"; Sylvester interview; Mayer, "Monfort Is a One-Company Industry."

54. "Corn Harvest," *Monfort of Colorado*.

55. Mayer, "Monfort Is a One-Company Industry." Statistics for corn grain and corn silage production in Colorado were produced in three respective census volumes and taken from U.S. Department of Commerce, *Agriculture Census of the United States: Colorado* (Washington: GPO, 1964, 1969, 1974).

56. Guinn, "Matsushima: Genius of the Feedlots"; "Production of Monfort"; Mayer, "Monfort Is a One-Company Industry"; Roe, "They Feed 200,000 Steers a Day"; Matsushima, *Broad Horizon*, 66–67. The majority of feeders did not possess the economies of scale to use Matsushima's flaked corn. Instead, they emphasized corn with a high moisture content that approximated the moisture content in the steam-flaking process.

57. "Packing Plant Slide Show," Monfort Collection, box 6, Scripts; Kenneth Monfort, interview by David McComb, November 16, 1973, Oral History of Colorado Project; Mayer, "Monfort Is a One Company Industry."

58. "Fabricating Meat at Monfort"; "Packing Plant Slide Show"; Monfort interview; Mayer, "Monfort Is a One-Company Industry."

59. Duane Flack, quoted in "Packaging Panorama," pamphlet produced by DOW Corporation, in Monfort Collection, box 7, folder 1; "Fabricating Meat at Monfort."

60. Richard Seaworth, owner, Seaworth Farms, interview by author, June 21, 2019, Wellington CO.

61. "MOPAC from Monfort of Colorado: MOPAC Retail Boxed Beef and Lamb Program," Monfort Collection, box 6, folders 8–11; "Packaging Panorama"; "Packing Plant Slide Show."

62. Monfort interview.

63. Piedmont farmers and feeders Don Ament, Richard Seaworth, Rod Ulrich, and Jerry Sonnenberg all confirm the transformation wrought by atrazine. Each spoke of its ability to kill all unwanted broadleaf weeds with residual effects that could carry over for several seasons. According to Seaworth, once a corn crop had been sprayed with atrazine "for the next four years you could not grow anything but corn." Ament interview; Seaworth interview; Rod Ulrich, cattle feeder, interview by author, LaSalle CO, June 27, 2019; Jerry Sonnenberg, June 24, 2019, interview by author.

64. W. D. Farr, "Speech at Iowa Cattle Feeder's Day."

65. On the place of organisms as technology, see Russell, "Can Organisms Be Technology?" 249–64.

PERSPECTIVE

1. Kent Peppler, past president of the Rocky Mountain Farmers' Union, interview by author, June 18, 2019, Mead CO.

2. Peppler interview.

3. For a fuller description of fracking, including its water requirements and political economy, see Ladd, *Fractured Communities*.

4. Peppler interview; Ryan Maye Handy, "Kent Peppler, 4th Generation Mead Farmer, Is Here to Stay," *Coloradoan*, September 26, 2014.

5. Stoll, *Fruits of Natural Advantage*, xii–xvi.

6. While Rachel Carson produced the most widely read critique of pesticides among ecologists during the period, she was by no means the only one. See for example Carson, *Silent Spring*; Rudd, *Pesticides and the Living Landscape*; Bookchin, *Our Synthetic Environment*; Commoner, *Science and Survival*.

7. Duncan, *Centrality of Agriculture*, 107–26. The work of entomologist Robert Van Den Bosch on integrated pest management provides a notable exception. See Van Den Bosch, *Pesticide Conspiracy*.

8. Robert Zimdahl, professor emeritus of weed science at Colorado State University, interview by author, June 14, 2016, Fort Collins CO.

9. Zimdahl interview.

10. Zimdahl interview.

11. Zimdahl interview.

12. The other half of Agent Orange is 2,4-D, which is still among the most common herbicides.

13. Robert Zimdahl, "Pesticides—A Value Question."

14. Zimdahl interview.

15. To find these statistics, I researched Colorado Agricultural statistics bulletins from 1890–1965. See Colorado Co-operative Crop Reporting Service et al., *Colorado Agricultural Statistics*.

16. Statistics cited in John K. Matsushima, *A Journey Back*; Bill Hamerich, CEO, Colorado Livestock Association, interview by author, Greeley CO, June 20, 2019.

17. Fitzgerald, *Every Farm a Factory*, 184–90.

18. Berry, *Unsettling of America*, 171–223.

19. Worster, "Good Farming and the Public Good," 31–42.

BIBLIOGRAPHY

ARCHIVES AND MANUSCRIPT MATERIALS

AEXT. Records of the Colorado Cooperative Extension. Colorado Agricultural and Natural Resource Archive, Morgan Library, Colorado State University, Fort Collins CO.

CCFA. Records of the Colorado Cattle Feeders Association. Colorado Agricultural and Natural Resource Archive, Morgan Library, Colorado State University, Fort Collins CO.

Civil Works Administration Pioneer Interviews. Stephen H. Hart Library, History Colorado, Denver.

Fort Morgan Museum, Fort Morgan CO.

GWS. Great Western Sugar Collection. Loveland Museum Archives, Loveland CO.

Hounshell and Smith Oral Histories. Hagley Library, Wilmington DE.

Irrigation Research Papers. Water Resources Archive, Morgan Library, Colorado State University, Fort Collins CO.

James Sanks Brisbin Papers, 1850–1891. Montana Historical Society, Helena MT.

JBS. Five Rivers Collection. City of Greeley Museum Permanent Collection, Greeley CO.

Jim Svedman Papers. Colorado Agricultural and Natural Resource Archive, Colorado State University, Fort Collins CO.

John Matsushima interview by James Hansen, March 2, 2010, transcript. Archives and Special Collections, Morgan Library, Colorado State University.

Loveland Historical Society Museum, Loveland CO.

NAL Sugar Beet Collection Reports. Special Collections. Records of the National Agricultural Library, Beltsville MD.

Oral History of Colorado Project. Stephen H. Hart Library, History Colorado, Denver.

Philip H. Boothroyd Collection. Norlin Library, University of Colorado, Boulder.

Ralph Parshall Papers. Water Resources Archive, Morgan Library, Colorado State University, Fort Collins CO.

Records of the Colorado Cattlemen's Association. Colorado Agricultural and Natural Resource Archive, Morgan Library, Colorado State University, Fort Collins CO.

Records of the Consolidated Home Supply Ditch and Reservoir Company. Water Resources Archive, Morgan Library, Colorado State University, Fort Collins CO.

Records of the Great Western Sugar Company, Agricultural and Natural Resources Archive, Colorado State University, Fort Collins CO.

Records of the Office of the Secretary of Agriculture. Records of the United States Department of Agriculture, National Archives, College Park MD.

Records of the Rocky Mountain Conference of Entomologists, Colorado Agricultural Archive, Colorado State University, Fort Collins CO.

Records of the United States Food Administration, National Archives, Denver CO.

RG 54. Records of the Bureau of Plant Industry, Soils, and Agricultural Engineering. Record Group 54. National Archives and Records Administration, College Park, MD.

SPI. Records of the Division of Sugar Plant Investigations. Records of the United States Department of Agriculture, National Archives, College Park MD.

TFM. Thomas F. Mahony Papers. Notre Dame Archives, South Bend IN.

W. D. Farr Collection. Water Resources Archive, Morgan Library, Colorado State University, Fort Collins CO.

Western Range Cattle Industry Study. Stephen H. Hart Library, History Colorado, Denver.

PUBLISHED WORKS

1957 Disease and Insect Control Recommendations for Colorado Vegetable Crop Insects. General Series Paper. Fort Collins: Colorado Agricultural Experiment Station, Colorado State University, 1957.

1961 Disease and Insect Control Recommendations for Colorado Vegetable Crop Insects. General Series Paper. Fort Collins: Colorado Agricultural Experiment Station, Colorado State University, 1961.

Ahlgren, Gilbert H., Glenn C. Klingman, and Dale E. Wolf. *Principles of Weed Control.* New York: Wiley, 1951.

Anderson, Esther Sanfreida. "Geography of the Beet Sugar Industry." Master's thesis, University of Nebraska, 1917.

Anderson, J. L. *Industrializing the Corn Belt: Agriculture, Technology, and Environment, 1945–1972.* DeKalb: Northern Illinois University Press, 2009.

Andrews, Thomas G. *Killing for Coal: America's Deadliest Labor War*. Cambridge MA: Harvard University Press, 2010.

Arrington, Leonard J. *Beet Sugar in the West: A History of the Utah-Idaho Sugar Company, 1891–1966*. Seattle: University of Washington Press, 1966.

———. "Science, Government, and Enterprise in Economic Development: The Western Beet Sugar Industry." *Agricultural History* 41, no. 1 (1967): 1–18.

Autobee, Robert. *Colorado-Big Thompson Project*. Denver: Bureau of Reclamation, 1996.

Baker, F. E. "Recollections of the Union Colony at Greeley." *Colorado Magazine*, July 1940.

Baker, O. E. *A Graphic Summary of American Agriculture, Based Largely on the Census of 1920*. Washington DC: U.S. Department of Agriculture, 1922.

Balderrama, Francisco E., and Raymond Rodriguez. *Decade of Betrayal: Mexican Repatriation in the 1930s*. Rev. ed. Albuquerque: University of New Mexico Press, 2006.

Ballinger, Roy A. *A History of Sugar Marketing through 1974*. Agricultural Economic Report, no. 382. Washington DC: Department of Agriculture, Economics, Statistics, and Cooperatives Service, 1978.

———. *The Sugar Industry and the Federal Government; a Thirty-Year Record, 1917–1947*. Washington DC: Sugar Statistics Service, 1949.

Barnhart, Walt. *Kenny's Shoes: A Walk through the Storied Life of the Remarkable Kenneth W. Monfort*. West Conshohocken PA: Infinity Publishing, 2008.

Bean, Geraldine B. *Charles Boettcher: A Study in Pioneer Western Enterprise*. Boulder CO: Westview Press, n.d.

Benton-Cohen, Katherine. *Borderline Americans: Racial Division and Labor War in the Arizona Borderlands*. Cambridge MA: Harvard University Press, 2009.

Bernhardt, Joshua. *Government Control of the Sugar Industry in the United States: An Account of the Work of the United States Food Administration and the United States Sugar Equalization Board, Inc*. New York: Macmillan, 1920.

Berry, Wendell. *The Unsettling of America: Culture and Agriculture*. San Francisco: Sierra Club Books, 1977.

Biancardi, Enrico, Leonard W. Panella, and Robert T. Lewellen. *Beta Maritima: The Origin of Beets*. New York: Springer, 2012.

Bookchin, Murray. *Our Synthetic Environment*. New York: Knopf, 1962.

Boyd, David. *A History: Greeley and the Union Colony of Colorado*. Greeley CO: Greeley Tribune Press, 1890.

Boyd, William. "Making Meat: Science, Technology, and American Poultry Production." *Technology and Culture* 42 (2001): 631–64.

Brandon, J. F., J. J. Curtis, and Warren Leonard. *Corn Production in Colorado.* Colorado Agricultural College Bulletin no. 463. Fort Collins: Colorado Agricultural College, 1940.

Bray, Charles I., and Colorado Agricultural Experiment Station. *Financing the Western Cattleman.* Bulletin no. 338. Fort Collins: Colorado Experiment Station, Colorado Agricultural College, 1928.

Brewbaker, H. E. "Studies of Irrigation Methods for Sugar Beets in Northern Colorado." *Journal of the American Society of Agronomy* 26, no. 3(1934).

Brisbin, James S., and Newberry Library. *The Beef Bonanza, or, How to Get Rich on the Plains Being a Description of Cattle-Growing, Sheep-Farming, Horse-Raising, and Dairying in the West.* Philadelphia: J. B. Lippincott, 1881.

Brown, Sara A., Robie O. Sargent, and Clara B. Armentrout. *Children Working in the Sugar Beet Fields of Certain Districts of the South Platte Valley, Colorado.* National Child Labor Committee. Publication no. 333. New York: National Child Labor Committee, 1925.

Buckmaster, J. C., and J. J Willis. *The Elementary Principles of Scientific Agriculture.* London: Moffatt and Paige, 1891.

Burdick, R. T., and H. B. Pingrey. *Profits from Winter Feeding in Northern Colorado.* Bulletin no. 394. Fort Collins: Agricultural Experiment Station, Colorado Agricultural College, 1932.

Bussing, Charles Earl. "The Cattle Feeding Industry of the Northern Colorado Piedmont: A Geographic Appraisal." Ph.D. diss., University of Nebraska, 1968.

Campbell, Larry, and A. Cattanach. "The American Society of Sugar Beet Technologists, Advancing Sugarbeet Research for 75 Years." *Journal of Sugar Beet Research* 50 (July 2013): 14–25.

Capozzola, Christopher Joseph Nicodemus. *Uncle Sam Wants You: World War I and the Making of the Modern American Citizen.* New York: Oxford University Press, 2008.

Carlyle, W. L. *Feeding Steers on Sugar Beet Pulp, Alfalfa Hay and Farm Grains.* Bulletin no. 97. Fort Collins: Agricultural Experiment Station, Agricultural College of Colorado, 1905.

Carpenter, Daniel P. *The Forging of Bureaucratic Autonomy: Reputations, Networks, and Policy Innovation in Executive Agencies, 1862–1928.* Princeton NJ: Princeton University Press, 2001.

Carson, Rachel. *Silent Spring.* Boston: Houghton Mifflin, 1962.

Chandler, Alfred D. *Shaping the Industrial Century: The Remarkable Story of the Modern Chemical and Pharmaceutical Industries.* Cambridge MA: Harvard University Press, 2005.

Clark, J. Maxwell. *Colonial Days.* Denver: Smith-Brooks, 1902.

Clemen, Rudolf Alexander. *The American Livestock and Meat Industry*. History of American Economy: Studies and Materials for Study. New York: Johnson Reprint, 1966.

Clopper, Edward Nicholas, and Lewis Wickes Hine. *Child Labor in the Sugar-Beet Fields of Colorado*. New York: National Child Labor Committee, 1916.

Cochrane, Willard Wesley. *The Development of American Agriculture: A Historical Analysis*. Minneapolis: University of Minnesota Press, 1979.

Coen, B. F. *Children Working on Farms in Certain Sections of Northern Colorado, Including Districts in the Vicinity of Windsor, Wellington, Fort Collins, Loveland, Longmont: Based Upon Studies Made During Summer, Fall and Winter, 1924, in Cooperation with National Child Labor Committee*. Colorado Agricultural College Bulletin, ser. 27, no. 2. Fort Collins: Colorado Agricultural College, 1926.

Colorado Agricultural Experiment Station, ed. *Alfalfa: Its Growth, Composition, Digestibility, Etc*. Bulletin no. 8. Fort Collins: Agricultural Experiment Station, Colorado Agricultural College, 1889.

Colorado Agricultural Statistics. Vols. for 1925–34 issued as series: Bulletin nos. 75, 81, 85–92. Denver: Colorado Co-operative Crop and Livestock Reporting Service, 1884.

Colorado, County Court (Weld County), and Juvenile Department. *The Farm and the School: A Resumé of a Survey of the Public Schools of Weld County, Colorado*. Greeley CO: Issued by the Extension Department of Colorado State Teachers College, 1918.

Colorado: Report on the Farmers' Costs of Producing Sugar Beets in Colorado; 1921, 1922, and 1923. Washington DC: GPO, 1926.

Commoner, Barry. *The Closing Circle: Nature, Man, and Technology*. 1st ed. New York: Knopf, 1972.

———. *Science and Survival*. New York: Viking Press, 1966.

Conkin, Paul Keith. *A Revolution down on the Farm: The Transformation of American Agriculture since 1929*. Lexington: University Press of Kentucky, 2009.

Cooke, Wells W. *Cattle Feeding in Colorado*. Bulletin no. 34. Fort Collins: Agricultural Experiment Station, Colorado Agricultural College, 1896.

Cooke, Wells W., and William Headden. *Sugar Beets in Colorado in 1897*. Bulletin no. 42. Fort Collins: Agricultural Experiment Station, Colorado Agricultural College, 1898.

Coons, George H. "The Sugar Beet: Product of Science." *Scientific Monthly* 68, no. 3 (1949): 149–64.

"Corn Harvest: Employees Harvest 250,000 Tons of Feed during Annual Event," *Monfort of Colorado*, (Fall 1977).

Crisler, Carney Clark. "The Mexican Bracero Program with Special Reference to Colorado." Master's thesis, University of Denver, 1968.

Cronon, William. *Nature's Metropolis: Chicago and the Great West*. New York: W. W. Norton, 1991.

Cronon, William, George A. Miles, and Jay Gitlin, eds. *Under an Open Sky: Rethinking America's Western Past*. New York: W. W. Norton, 1992.

Crumpacker, David Wilson. *Performance Tests of Hybrid Corn Varieties Grown in Various Regions of Colorado in 1959*. Fort Collins: Agricultural Experiment Station, Colorado State University, 1960.

Cuff, Ray. "Livestock Pest Control," *Journal of Agricultural Chemicals* 9, no. 6 (1954).

Cullather, Nick. "The Foreign Policy of the Calorie." *American Historical Review* 112, no. 2 (2007): 337–64.

Cunfer, Geoff. *On the Great Plains: Agriculture and Environment*. College Station: Texas A&M University Press, 2005.

Cushman, Gregory T. *Guano and the Opening of the Pacific World: A Global Ecological History*. Studies in Environment and History. Cambridge: Cambridge University Press, 2013.

Dale, Edward Everett. *The Range Cattle Industry: Ranching on the Great Plains from 1865 to 1925*. New ed. Norman: University of Oklahoma Press, 1960.

Dalton, John Edward. *Sugar: A Case Study of Government Control*. New York: Macmillan, 1937.

Davan, Clarence Floyd, W. R. Schmehl, and W. G. Stewart. *Fertilizer Use and Trends for Principal Crops in Colorado*. Fort Collins: Colorado Agricultural Experiment Station, Colorado State University, 1962.

Davis, Frederick Rowe. *Banned: A History of Pesticides and the Science of Toxicology*. New Haven CT: Yale University Press, 2014.

Decker, Peter R. *"The Utes Must Go!": American Expansion and the Removal of a People*. Golden CO: Fulcrum, 2004.

De Ong, E. R. *Chemistry and Uses of Insecticides*. New York: Reinhold, 1948.

Deutsch, Sarah. *No Separate Refuge: Culture, Class, and Gender on an Anglo-Hispanic Frontier in the American Southwest, 1880–1940*. New York: Oxford University Press, 1987.

Donahue, Brian. *The Great Meadow: Farmers and the Land in Colonial Concord*. Yale Agrarian Studies. New Haven CT: Yale University Press, 2004.

Donato, Rubén. *Mexicans and Hispanos in Colorado Schools and Communities, 1920–1960*. Albany: State University of New York Press, 2007.

Dove, W. E. "Livestock Insect Control." *Journal of Agricultural Chemicals* 1, no. 1 (1946).

Duffin, Andrew P. *Plowed Under: Agriculture and Environment in the Palouse.* Seattle: University of Washington Press, 2007.

Duncan, Colin A. M. *The Centrality of Agriculture: Between Humankind and the Rest of Nature.* Montreal: McGill-Queen's University Press, 1996.

Dunlap, Thomas R. *DDT: Scientists, Citizens, and Public Policy.* Princeton NJ: Princeton University Press, 1981.

Dupree, A. Hunter. *Science in the Federal Government: A History of Policies and Activities to 1940.* Cambridge MA: Belknap Press of Harvard University Press, 1957.

Eddy, Edward Danforth. *Colleges for Our Land and Time: The Land-Grant Idea in American Education.* Westport CT: Greenwood Press, 1973.

"Elbridge Gerry, Colorado Pioneer." *Colorado Magazine*, April 1952: 137–49.

Ellis, Lippert S., John R. Commons, Benjamin Horace Hibbard, and Walter A. Morton. *The Tariff on Sugar.* Freeport IL: Rawleigh Foundation, 1933.

Elmore, Bartow J. *Citizen Coke: The Making of Coca-Cola Capitalism.* First edition. New York: W. W. Norton, 2015.

"Fabricating Meat at Monfort: From Kill to Fill." *National Provisioner*, May 17, 1969.

Faurot, C. S. *A Practical Talk to Practical Farmers on Sugar Beet Culture.* Longmont CO: Longmont Call Press, 1903.

Fiege, Mark. *Irrigated Eden: The Making of an Agricultural Landscape in the American West.* Seattle: University of Washington Press, 1999.

Fite, Gilbert Courtland. *The Farmers' Frontier, 1865–1900.* New York: Holt, Rinehart, and Winston, 2002.

Fitzgerald, Deborah Kay. *The Business of Breeding: Hybrid Corn in Illinois, 1890–1940.* Ithaca: Cornell University Press, 1990.

———. *Every Farm a Factory: The Industrial Ideal in American Agriculture.* Yale Agrarian Studies Series. New Haven CT: Yale University Press, 2003.

"Flies Make Profits Fly." *Down to Earth* 3, no. 2 (1947).

Flores, Dan L. *The Natural West: Environmental History in the Great Plains and Rocky Mountains.* Norman: University of Oklahoma Press, 2001.

Foley, Neil. *The White Scourge: Mexicans, Blacks, and Poor Whites in Texas Cotton Culture.* Vol. 2 of American Crossroads. Berkeley: University of California Press, 1997.

Folks, Homer, and National Child Labor Committee, eds. *Changes and Trends in Child Labor and Its Control.* New York: National Child Labor Committee, 1938.

Fortier, Samuel. *Irrigation Requirements of the Arid and Semiarid Lands of the Missouri and Arkansas River Basins.* USDA Bulletin no. 36. Washington DC: GPO, 1928.

Frank, Jerry J. *Making Rocky Mountain National Park: The Environmental History of an American Treasure*. Lawrence: University Press of Kansas, 2013.

Freeborn, Stanley. "How Experiment Stations Help Develop New Pesticides." *Journal of Agricultural Chemicals* 3, no. 9 (1948): 1.

French, Jasper J., and Herman Fauber. *Tests of Hybrid Corn under Irrigation in Colorado in 1945*. Fort Collins: Colorado Agricultural Experiment Station, Colorado State College, 1946.

Frink, Maurice, W. Turrentine Jackson, and Agnes Wright Spring. *When Grass Was King: Contributions to the Western Range Cattle Industry Study*. Boulder: University of Colorado Press, 1956.

Gardner, Robert Alexander, and D. W. Robertson. *The Nitrogen Requirement of Sugar Beets*. Technical Bulletin 28. Fort Collins: Colorado Agricultural Experiment Station, Colorado State College, 1942.

Gardner, Robert Alexander, D. W. Robertson, and Rodney Tucker. *Looking at the 1947 Soil Fertility Problem as It Affects Sugar Beet Production*. Miscellaneous Series Paper. Fort Collins: Agricultural Experiment Station, Colorado State College, 1947.

Gausman, Glen J., and Herman Fauber. *Tests of Hybrid Corn under Irrigation in Colorado 1948*. Fort Collins: Agricultural Experiment Station, Colorado State College, 1948.

Gilbert, Jess Carr. *Planning Democracy: Agrarian Intellectuals and the Intended New Deal*. Yale Agrarian Studies Series. New Haven CT: Yale University Press, 2015.

Godfrey, Matthew C. *Religion, Politics, and Sugar: The Mormon Church, the Federal Government, and the Utah-Idaho Sugar Company, 1907–1921*. Logan: Utah State University Press, 2007.

Goff, Richard. *Century in the Saddle*. Colorado Cattlemen's Centennial Commission, 1967.

Gonzalez, Gilbert. "Mexican Labor Migration, 1876–1924." In *Beyond La Frontera: The History of Mexico-U.S. Migration*, edited by Mark Overmyer-Velázquez. New York: Oxford University Press, 2011.

Goodman, James. *Stories of Scottsboro*. New York: Pantheon Books, 1994.

Greeley, Horace, and Robert Dale Owen. *The Autobiography of Horace Greeley: Or, Recollections of a Busy Life: To Which Are Added Miscellaneous Essays and Papers*. New York: E. B. Treat, 1981.

———. *What I Know of Farming: A Series of Brief and Plain Expositions of Practical Agriculture as an Art Based upon Science*. New York: Tribune, 1871.

Guinn, Jack. "John Matsushima: Genius of the Feedlots." *Empire Magazine*, August 13, 1967: 14–17.

Gutiérrez, David. *Walls and Mirrors: Mexican Americans, Mexican Immigrants and the Politics of Ethnicity*. Berkeley: University of California Press, 1995.

Ham, William Thomas. "Regulation of Labour Conditions in Sugar Cultivation under the Agricultural Adjustment Act." *International Labour Review* 33 (1936).

Hamalainen, Pekka. *The Comanche Empire*. New Haven CT: Yale University Press, 2009.

Hamilton, Shane. *Trucking Country: The Road to America's Wal-Mart Economy*. Politics and Society in Modern America. Princeton NJ: Princeton University Press, 2008.

Hansen, James E. *Beyond the Ivory Tower: A History of Colorado State University Cooperative Extension*. Colorado State University Cooperative Extension, n.d.

Hart, Elva Treviño. *Barefoot Heart: Stories of a Migrant Child*. Tempe AZ: Bilingual Press/Editorial Bilingüe, 2006.

Hart, John Fraser. *The Changing Scale of American Agriculture*. Charlottesville: University of Virginia Press, 2003.

Hays, Willet. "Progress in Plant and Animal Breeding." In *Yearbook of the United States Department of Agriculture, 1901*. United States Department of Agriculture. Washington DC: GPO, 1902.

Haynes, Emma S. "German-Russians on the Volga and in the United States." Master's thesis, University of Oregon, 1929.

Headden, W. P. *Deterioration in the Quality of Sugar Beets Due to Nitrates Formed in the Soil*. Fort Collins: Agricultural Experiment Station, Colorado Agricultural College, 1912.

——. *The Fixation of Nitrogen in Some Colorado Soils*. Fort Collins: Agricultural Experiment Station, Colorado Agricultural College, 1910.

Hemphill, Robert G., and United States Department of Agriculture, eds. *Irrigation in Northern Colorado*. USDA Bulletin no. 1026. Washington DC: GPO, 1922.

Henderson, David Allen. "The Beef Cattle Industry of Colorado." PhD diss., University of Colorado, Boulder, 1951.

Henry, William, and Frank Morrison. *Feeds and Feeding: A Handbook for the Student and Stockman . . . Seventeenth Edition*. Madison WI: Henry-Morrison, 1917.

"The Herbicidal Action of 2,4 Dichlorophenoxyacetic and 2,4,5 Trichlorophenoxyacetic Acid on Bindweeds." *Science*, August 18, 1944.

Hernandez, Elvira C., and Eduardo Hernandez-Chavez. *Elvira: A Mexican Immigrant Woman*. Stockton CA: Ediciones Lengua y Cultura, 2017.

Heston, Thomas J. *Sweet Subsidy: The Economic and Diplomatic Effects of the U.S. Sugar Acts, 1934–1974*. New York: Garland, 1987.

Higham, John. *Strangers in the Land: Patterns of American Nativism, 1860–1925*. New Brunswick NJ: Rutgers University Press, 2002.

Hightower, Jim, and Agribusiness Accountability Project. *Hard Tomatoes, Hard Times: The Original Hightower Report, Unexpurgated, of the Agribusiness Accountability Project on the Failure of America's Land Grant College Complex and Selected Additional Views of the Problems and Prospects of American Agriculture in the Late Seventies*. Cambridge MA: Schenkman, 1978.

Hobbs, Greg. *The Public's Water Resource: Articles on Water Law, History, and Culture*. 2nd ed. Denver: Continuing Legal Education in Colorado, 2010.

Hoganson, Kristin L. *Fighting for American Manhood: How Gender Politics Provoked the Spanish-American and Philippine-American Wars*. New Haven CT: Yale University Press, 2000.

Hundley, Norris. *Water and the West: The Colorado River Compact and the Politics of Water in the American West*. 2nd ed. Berkeley: University of California Press, 2009.

Hurt, R. Douglas. *Agricultural Technology in the Twentieth Century*. Manhattan KS: Sunflower University Press, 1991.

Hutchins, Wells A. *Mutual Irrigation Companies*. USDA Bulletin no. 82. Washington DC: GPO, 1929.

Isenberg, Andrew C. *The Destruction of the Bison: An Environmental History, 1750–1920*. Studies in Environment and History. New York: Cambridge University Press, 2000.

Jackson, Wes. *Altars of Unhewn Stone: Science and the Earth*. San Francisco: North Point Press, 1987.

Jacobson, Matthew Frye. *Whiteness of a Different Color: European Immigrants and the Alchemy of Race*. Cambridge MA: Harvard University Press, 1998.

Jenkins, Tom. "Hardworking John Wesley Iliff Became the Most Prominent Cattle Baron in Colorado." *Wild West*, October 2004.

Jennings, R. D. *Consumption of Feed by Livestock, 1909–47: Relation between Feed, Livestock, and Food at the National Level*. USDA Bulletin no. 836. Washington DC: GPO, 1949.

Jenny, Hans. "The Making and Unmaking of a Fertile Soil." In *Meeting the Expectations of the Land: Essays in Sustainable Agriculture and Stewardship*, edited by Wes Jackson, Wendell Berry, and Bruce Colman, 42–55. San Francisco: North Point Press, 1984.

Jessen, Kenneth Christian. *Railroads of Northern Colorado*. 1st ed. Boulder CO: Pruett, 1982.

Johnson, Elizabeth S. *Wages, Employment Conditions, and Welfare of Sugar-Beet Laborers*. Washington DC: U.S. Department of Labor. Bureau of Labor Statistics, 1938.

———. *Welfare of Families of Sugar-Beet Laborers: A Study of Child Labor and Its Relation to Family Work, Income, and Living Conditions in 1935.* United States Children's Bureau. Bureau Publication, no. 247. Washington DC: GPO, 1939.

Keirnes, W. R. *Water: Colorado's Most Precious Asset: The Consolidated Home Supply Ditch and Reservoir Company Contributed to Its Development, 1881–1986.* Loveland CO: Mile-High Printing, 1986.

Kelley, Robin D. G. *Hammer and Hoe: Alabama Communists during the Great Depression.* Fred W. Morrison Series in Southern Studies. Chapel Hill: University of North Carolina Press, 1990.

Kezer, Alvin. *Irrigated Farming in Colorado: Being an Authoritative Statement of Facts about this Method of Farming in Colorado.* Denver: Colorado State Board of Immigration, 1900.

Kingsolver, Barbara, Steven L. Hopp, and Camille Kingsolver. *Animal, Vegetable, Miracle: A Year of Food Life.* 1st ed. New York: HarperCollins, 2007.

Kloberdanz, Timothy J. *The Volga Germans in Old Russia and in Western North America: Their Changing World View.* Lincoln NE: American Historical Society of Germans from Russia, 1979.

Kloppenburg, Jack Ralph. *First the Seed: The Political Economy of Plant Biotechnology, 1492–2000.* 2nd ed. Science and Technology in Society. Madison: University of Wisconsin Press, 2004.

Knipling, E. F. "Arthur W. Lindquist, 1903–1980." *Bulletin of the Entomological Society of America* 26, no. 2 (June 1980): 135–36.

Koch, Fred C. *The Volga Germans: In Russia and the Americas, from 1763 to the Present.* University Park: Pennsylvania State University Press, 1977.

Kramer, Paul A. *The Blood of Government: Race, Empire, the United States, and the Philippines.* Chapel Hill: University of North Carolina Press, 2006.

Ladd, Anthony E., ed. *Fractured Communities: Risk, Impacts, and Protest against Hydraulic Fracking in U.S. Shale Regions.* New Brunswick NJ: Rutgers University Press, 2018.

Laflin, Rose. *Irrigation, Settlement, and Change on the Cache La Poudre River.* Colorado Water Resources Research Institute, Colorado State University, n.d.

Langston, Nancy. *Toxic Bodies: Hormone Disruptors and the Legacy of DES.* New Haven CT: Yale University Press, 2010.

Larson, Olaf F. *Beet Workers on Relief in Weld County, Colorado.* Research Bulletin no. 4. Fort Collins: Agricultural Experiment Station, Colorado State College, 1937.

Leigh, G. J. *The World's Greatest Fix: A History of Nitrogen and Agriculture.* Oxford: Oxford University Press, 2004.

Leonard, Warren H., Herman Fauber, and R. H. Tucker. *Tests with Hybrid Corn in Colorado*. Fort Collins: Colorado Agricultural Experiment Station, Colorado State College, 1940.

Lewis, C. S. *The Abolition of Man or Reflections on Education with Special Reference to the Teaching of English in the Upper Forms of Schools*. New York: Macmillan, 1947.

Lindenmeyer, Kriste. *A Right to Childhood: The U.S. Children's Bureau and Child Welfare, 1912–46*. Urbana: University of Illinois Press, 1997.

Lindquist, Arthur. "Livestock Insect Control." *Journal of Agricultural Chemicals* 15, no. 11 (1960).

Lopez, Jody L., Gabriel A. Lopez, and Peggy A. Ford. *White Gold Laborers: The Spanish Colony of Greeley, Colorado*. Bloomington IN: Author House, 2007.

Maass, Arthur, and Raymond Lloyd Anderson. *And the Desert Shall Rejoice: Conflict, Growth, and Justice in Arid Environments*. Cambridge MA: MIT Press, 1978.

Macy, Loring K., Lloyd E. Arnold, Eugene G. McKibben, and Edmund J. Stone. *Changes in Technology and Labor Requirements in Crop Production: Sugar Beets*. Philadelphia: Works Progress Administration, 1937.

Madden, J. Patrick, and Elmer C. Hunter. *Economies of Size for Specialized Beef Feedlots in Colorado*, Washington DC: U.S. Department of Agriculture, 1966.

Madsen, Albert Gail. *Response of Colorado Cattle Feeders to Selected Economic Factors 1961–1969*. Fort Collins: Colorado Agricultural Experiment Station, Colorado State University, 1971.

Magnuson, Torsten A. "History of the Beet Sugar Industry in California." *Annual Publication of the Historical Society of Southern California* 11, no. 1 (1918): 68–79.

Mahony, Thomas F. *Problem of the Mexican Wage Earner*. Denver: Knights of Columbus, 1930.

Mangelsdorf, Paul C. "Hybrid Corn," *Scientific American* 185, no. 2 (1951): 39–47.

Mapes, Kathleen. *Sweet Tyranny: Migrant Labor, Industrial Agriculture, and Imperial Politics*. The Working Class in American History. Urbana: University of Illinois Press, 2009.

Marcus, Alan I. *Agricultural Science and the Quest for Legitimacy: Farmers, Agricultural Colleges, and Experiment Stations,1870–1890*. The Henry A. Wallace Series on Agricultural History and Rural Studies. Ames: Iowa State University Press, 1985.

——— . *Cancer from Beef: DES, Federal Food Regulation, and Consumer Confidence*. Baltimore MD: Johns Hopkins University Press, 1994.

——— . *Service as Mandate: How American Land-Grant Universities Shaped the Modern World, 1920–2015*. Tuscaloosa: University Alabama Press, 2015.

Mart, Michelle. *Pesticides, A Love Story: America's Enduring Embrace of Dangerous Chemicals*. Lawrence: University Press of Kansas, 2018.

Matsushima, John K. *Broad Horizon: I Fear No Boundaries: An Autobiography*. North Charleston SC: CreateSpace, 2011.

———. *A Journey Back: A History of Cattle Feeding in Colorado and the United States*. Colorado Springs: Cattlemen's Communications, 1995.

Matthews, Ellen Nathalie, Gertrude Underhill Light, eds. *Child Labor and the Work of Mothers in the Beet Fields of Colorado and Michigan*. Bureau Publication, no. 115. Washington DC: U.S. Department of Labor, Children's Bureau: GPO, 1923.

Maxson, A. C. *Principal Insect Enemies of the Sugar Beet in the Territories Served by the Great Western Sugar Company*. Denver CO: Agricultural Department, Great Western Sugar Company, 1920.

May, William John. *The Great Western Sugarlands: The History of the Great Western Sugar Company and the Economic Development of the Great Plains*. Garland Studies in Entrepreneurship. New York: Garland, 1989.

Mayer, Lawrence. "Monfort Is a One-Company Industry." *Fortune*, January 1973.

Maynard, E. J. *Beet By-Products for Fattening Lambs*. Fort Collins: Agricultural Experiment Station, Colorado Agricultural College, 1921.

———. *Beets and Meat: A Practical Manual for Fattening Cattle and Sheep on Irrigated Farms of the Intermountain Area Where Sugar Beet By-Products Provide One of the Principal Sources of Low-Cost Fattening Feed*. Denver CO: Through the Leaves Press, 1945.

———. *Colorado Fattening Rations for Cattle, 1936*. Agricultural Experiment Station, Colorado State College, 1936.

———. *Feedlot Fattening Rations for Cattle: Progress Report of Livestock Feeding Experiment 1930*. Fort Collins: Agricultural Experiment Station, Colorado Agricultural College, 1930.

McClymonds, A. E. *Colorado Corn Problems*. Colorado Agricultural College Bulletin, ser. 1, no. 170-A. Fort Collins: Colorado Agricultural College, 1920.

McGerr, Michael. "Is There a Twentieth Century West?" In *Under an Open Sky: Rethinking America's Western Past*, edited by William Cronon, George A. Miles, and Jay Gitlin, 239–56. New York: W. W. Norton, 1992.

McWilliams, Carey. *Ill Fares the Land: Migrants and Migratory Labor in the United States*. New York: Arno Press, 1976.

McWilliams, James E. *American Pests: The Losing War on Insects from Colonial Times to DDT*. New York: Columbia University Press, 2008.

Mead, Elwood. *Irrigation Institutions: A Discussion of the Economic and Legal Questions Created by the Growth of Irrigated Agriculture in the West*. Use and Abuse of America's Natural Resources. New York: Arno Press, 1972.

Mehls, Carol Drake. *Weld County, Colorado Historic Agricultural Context.* Office of Archaeology and Historic Preservation, Colorado Historical Society, n.d.

Mehls, Steven. *The New Empire of the Rockies: A History of Northeast Colorado.* CreateSpace, 2014.

Merleaux, April. *Sugar and Civilization: American Empire and the Cultural Politics of Sweetness.* Chapel Hill: University of North Carolina Press, 2015.

Merrill, Karen R. "In Search of the 'Federal Presence' in the American West." *Western Historical Quarterly* 30, no. 4 (1999): 449–73.

Michener, James A. *Centennial.* New York: Random House, 1974.

Mintz, Sidney Wilfred. *Sweetness and Power: The Place of Sugar in Modern History.* New York: Viking, 1985.

"Monfort: The Legend." *Chilton's Commercial Car Journal,* September 1978: 101–10.

Moorhouse, L. A., R. S. Washburn, T. H. Summers, and S. B. Nuckols. *Farm Practice in Growing Sugar Beets for Three Districts in Colorado 1914–15.* USDA Bulletin no. 726. Washington DC: GPO, 1918.

Moorhouse, L. A., and T. H. Summers. *Saving Man Labor in Sugar-Beet Fields.* Washington DC: GPO, 1928.

Morgan, Gary. *Sugar Tramp: Colorado's Great Western Railway.* Fort Collins CO: Centennial Publications, 1975.

Morison, Elting E. *From Know-How to Nowhere: The Development of American Technology.* New York: Basic Books, 1974.

Muir, John. *My First Summer in the Sierra.* Boston: Houghton Mifflin, 1916.

Nash, Linda. "The Fruits of Ill-Health: Pesticides and Workers' Bodies in Post–World War II California." *Osiris* 19 (2004): 203.

Ngai, Mae M. *Impossible Subjects: Illegal Aliens and the Making of Modern America.* Politics and Society in Modern America. Princeton NJ: Princeton University Press, 2004.

Nuckols, S. B., United States Department of Agriculture, and Office of the Secretary. *Farm Practice in Growing Sugar Beets for 3 Districts in Colorado, 1914–15.* Washington DC: GPO, 1918.

Oleson, Alexandra, and John Voss, eds. *The Organization of Knowledge in Modern America, 1860–1920.* Baltimore MD: Johns Hopkins University Press, 1979.

Olmstead, Alan L. *Creating Abundance: Biological Innovation and American Agricultural Development.* New York: Cambridge University Press, 2008.

Parshall, Ralph. *Agricultural Economic Summary Relating to the Colorado-Big Thompson Project.* Fort Collins CO, publisher not identified 1937.

———. *Measuring Water in Irrigation Channels with Parshall Flumes and Small Weirs.* Washington DC: GPO, 1950.

———. *Parshall Flumes of Large Size*. Fort Collins: Colorado Agricultural Extension Service, 1954.

———. *The Parshall Measuring Flume*. Bulletin no. 423. Fort Collins: Agricultural Experiment Station, Colorado State College, 1936.

———. *Return of Seepage Water to the Lower South Platte River in Colorado*. Bulletin no. 279. Fort Collins: Agricultural Experiment Station, Colorado Agricultural College, 1922.

"Pauper Labor and Child Labor." *The Nation*. July 8, 1931.

Pearl, Raymond. "'Memorandum for Sugar Equalization Board Re: The Importance of Sugar as Food,' August 9, 1918, Statistical Division, U.S. Food Administration." In *Government Control of the Sugar Industry in the United States: An Account of the Work of the United States Food Administration and the United States Sugar Equalization Board, Inc.*, by Joshua Bernhardt, 230–32. New York: Macmillan, 1920.

Perkins, John H. *Insects, Experts, and the Insecticide Crisis: The Quest for New Pest Management Strategies*. New York: Plenum Press, 1982.

Pfaff, Christine. *The Colorado-Big Thompson Project Historic Context and Description of Property Types*. Denver: Bureau of Reclamation, 1999.

Pisani, Donald J. *From the Family Farm to Agribusiness: The Irrigation Crusade in California, 1850–1931*. Berkeley: University of California Press, 1984.

———. *To Reclaim a Divided West: Water, Law, and Public Policy,1848–1902*. 1st ed. Albuquerque: University of New Mexico Press, 1992.

———. *Water and American Government: The Reclamation Bureau, National Water Policy, and the West, 1902–1935*. Berkeley: University of California Press, 2002.

Pollan, Michael. *The Omnivore's Dilemma: A Natural History of Four Meals*. New York: Penguin Books, 2007.

"Portrait of a Packer-Feeder." *Meat*, August 1969.

Proceedings of the Agricultural Chemicals Dealers Conference, Panel on Insecticides. Denver, December 7, 1945. Fort Collins: Agricultural Experiment Station, Colorado Agricultural College, 1946.

Rasmussen, Wayne D. *Readings in the History of American Agriculture*. Urbana: University of Illinois Press, 1966.

———. "Technological Change in Western Sugar Beet Production." *Agricultural History* 41, no. 1 (1967): 31–36.

Rasmussen, Wayne D., and Gladys Baker. *The Department of Agriculture*. Praeger Library of U.S. Government Departments and Agencies, no. 32. New York: Praeger, 1972.

Richthofen, Walter von. *Cattle-Raising on the Plains of North America*. New York: Appleton, 1885.

Rock, Kenneth. "'Unsere Leute': The Germans from Russia in Colorado." *Colorado Magazine*, Spring 1977, 154–83.

Roe, Leonard. "They Feed 200,000 Steers a Day." *Monfort of Colorado*, (March/April 1972).

Romero, Adam M. "'From Oil Well to Farm': Industrial Waste, Shell Oil, and the Petrochemical Turn (1927–1947)." *Agricultural History* 90, no. 1 (2016): 70–93.

Rosenberg, Charles E. *No Other Gods: On Science and American Social Thought.* Revised and expanded ed. Baltimore MD: Johns Hopkins University Press, 1997.

Rossiter, Margaret W. *The Emergence of Agricultural Science: Justus Liebig and the Americans, 1840–1880.* New Haven CT: Yale University Press, 1975.

———. "The Organization of the Agricultural Sciences." In *The Organization of Knowledge in Modern America, 1860–1920*, edited by Alexandra Oleson and John Voss, 211–48. Baltimore MD: Johns Hopkins University Press, 1979.

Rudd, Robert L. *Pesticides and the Living Landscape.* A Conservation Foundation Study. Madison: University of Wisconsin Press, 1964.

Russell, Edmund. "Can Organisms Be Technology?" In *The Illusory Boundary: Environment and Technology in History*, edited by Martin Reuss and Stephen H. Cutcliffe, 249–64. Charlottesville: University of Virginia Press, 2010.

———. "The Strange Career of DDT: Experts, Federal Capacity, and Environmentalism in World War II." *Technology and Culture* 40, no. 4 (1999): 770–96.

———. *War and Nature: Fighting Humans and Insects with Chemicals from World War I to Silent Spring.* Studies in Environment and History. New York: Cambridge University Press, 2001.

Sabin, Dena Markoff. *The Colorado Beet Boom, 1899–1926: Growth and Development of the State's Sugar Industry.* Arvada CO: Western Heritage Conservation, 1981.

Sackman, Douglas C. "'Nature's Workshop': The Work Environment and Workers' Bodies in California's Citrus Industry, 1900–1940." *Environmental History* 5, no. 1 (2000): 27–53.

———. *Orange Empire: California and the Fruits of Eden.* Berkeley: University of California Press, 2007.

Sánchez, Saúl. *Rows of Memory: Journeys of a Migrant Sugar-Beet Worker.* Iowa City: University of Iowa Press, 2014.

Sanders, M. Elizabeth. *Roots of Reform: Farmers, Workers, and the American State, 1877–1917.* Chicago: University of Chicago Press, 1999.

Sawyer, Richard Clark. *To Make a Spotless Orange: Biological Control in California.* 1st ed. The Henry A. Wallace Series on Agricultural History and Rural Life. Ames: Iowa State University Press, 1996.

Saylor, Charles. *Progress of the Beet-Sugar Industry in the United States in 1901.* United States Congressional Serial Set no. 4241. Washington DC: GPO, 1902.

Schorr, David. *The Colorado Doctrine: Water Rights, Corporations, and Distributive Justice on the American Frontier.* Yale Law Library Series in Legal History and Reference. New Haven CT: Yale University Press, 2012.

Schwartz, Harry. *Seasonal Farm Labor in the United States: With Special Reference to Hired Workers in Fruit and Vegetable and Sugar-Beet Production.* Columbia University Studies in the History of American Agriculture Vol. 11. New York: Columbia University Press, 1945.

Scott, Roy Vernon. *The Reluctant Farmer: The Rise of Agricultural Extension to 1914.* Urbana: University of Illinois Press, 1971.

Scranton, Philip, and Susan R. Schrepfer, eds. *Industrializing Organisms: Introducing Evolutionary History.* Hagley Perspectives on Business and Culture Vol. 5. New York: Routledge, 2004.

Shaw, Chad Delano. "Twice Separated without a State: German-Russians in Weld County, 1900–1920." Master's thesis, University of Northern Colorado, 2006.

Sheflin, Douglas. *Legacies of Dust: Land Use and Labor on the Colorado Plains.* Lincoln: University of Nebraska Press, 2019.

Sherow, James Earl. *The Grasslands of the United States: An Environmental History.* ABC-CLIO's Nature and Human Societies Series. Santa Barbara CA: ABC-CLIO, 2007.

Skaggs, Jimmy M. *Prime Cut: Livestock Raising and Meatpacking in the United States, 1607–1983.* 1st ed. College Station: Texas A&M University Press, 1986.

Skuderna, A. W., ed. *Agronomic Evaluation Tests on Mechanical Blocking and Cross Cultivation of Sugar Beets.* USDA Technical Bulletin no. 316. Washington DC: GPO, 1934.

Smil, Vaclav. *Enriching the Earth: Fritz Haber, Carl Bosch, and the Transformation of World Food Production.* Cambridge MA: MIT Press, 2001.

Smythe, William E. *The Conquest of Arid America.* New and rev. ed. New York: Macmillan, 1905.

Snay, Mitchell. *Horace Greeley and the Politics of Reform in Nineteenth-Century America.* Lanham MD: Rowman and Littlefield, 2011.

Soluri, John. *Banana Cultures: Agriculture, Consumption, and Environmental Change in Honduras and the United States.* Austin: University of Texas Press, 2005.

Specht, Joshua. *Red Meat Republic: A Hoof-to-Table History of How Beef Changed America.* Princeton NJ: Princeton University Press, 2020.

Spencer, Guilford L. *Utilization of Residues from Beet-Sugar Manufacture in Cattle Feeding.* Washington DC: U.S. Department of Agriculture, 1899.

Standish, Sierra. "Beet Borderland: Hispanic Workers, the Sugar Beet, and the Making of a Northern Colorado Landscape." Master's thesis, Colorado State University, 2002.

Steinberg, Theodore. *Nature Incorporated: Industrialization and the Waters of New England.* Reprint ed. Amherst: University of Massachusetts Press, 1994.

Steinel, Alvin T., and Colorado Agricultural College. *History of Agriculture in Colorado: A Chronological Record of Progress in the Development of General Farming, Livestock Production, and Agricultural Education and Investigation on the Western Border of the Great Plains and in the Mountains of Colorado, 1858 to 1926.* Fort Collins: Colorado Agricultural College, 1926.

Steuter, Allen, and Lori Hidinger. "Comparative Ecology of Bison and Cattle on Mixed-Grass Prairie." *Great Plains Research: A Journal of Natural and Social Sciences* 9 (1999): 15.

Stewart, Mart A. *"What Nature Suffers to Groe": Life, Labor, and Landscape on the Georgia Coast, 1680–1920.* Athens: University of Georgia Press, 2002.

Stoll, Steven. *The Fruits of Natural Advantage: Making the Industrial Countryside in California.* Berkeley: University of California Press, 1998.

——. *Larding the Lean Earth: Soil and Society in Nineteenth-Century America.* New York: Hill and Wang, 2002.

Summons, Terry G. "Animal Feed Additives, 1940–1966." *Agricultural History* 42, no. 4: 305–13.Sutter, Paul, and Christopher J. Manganiello, eds. *Environmental History and the American South: A Reader.* Environmental History and the American South. Athens: University of Georgia Press, 2009.

Taylor, Paul S. *Mexican Labor in the United States Valley of the South Platte. Vol. 6, No. 2.* Berkeley: University of California Press, 1928.

Thornton, Bruce Jay, and Colorado Agricultural Experiment Station. *Report on Herbicides.* Fort Collins: Colorado Agricultural Experiment Station, Colorado State University, 1957.

Thornton, Bruce Jay, Jess Lafayette Fults, D. W. Robertson, and R. H. Tucker. *Panel on Weed Control.* Miscellaneous Series Paper. Fort Collins: Agricultural Experiment Station, Colorado State College, 1946.

Townsend, Charles. *The Beet-Sugar Industry in the United States in 1920.* USDA Technical Bulletin no. 995. Washington DC: GPO, 1921.

——. "Conditions Affecting the Production of Sugar Beet Seeds in the United States." *Yearbook of the United States Department of Agriculture, 1909.* United States Congressional Serial Set no. 5772. Washington DC: GPO, 1910.

Twitty, Eric. *The Silver Wedge: The Sugar Beet Industry in Fort Collins: A Historical Survey.* Westminster CO: SWCA Environmental Consultants, 2003.

Tyler, Daniel. *The Last Water Hole in the West: The Colorado-Big Thompson Project and the Northern Colorado Water Conservancy District.* Niwot: University Press of Colorado, 1992.

———. *WD Farr: Cowboy in the Boardroom.* Norman: University of Oklahoma Press, 2011.

Ubbelohde, Carl, Maxine Benson, and Duane A. Smith. *A Colorado History.* Boulder CO: Pruett, 2006.

Union Colony of Colorado. *First Annual Report of the Union Colony of Colorado, 1871: Including a History of the Town . . . of Greeley, from Its Date of Settlement to the Present.* New York: George W. Southwick, 1871.

United States Congress, House Committee on Immigration and Naturalization. *Immigration from Countries of the Western Hemisphere. Hearings . . . on H.R. 6465, 7358, 10955, 11687 . . . Feb. 21 to April 5, 1928.* Washington DC: GPO, 1928.

United States Congress, House. *Seasonal Agricultural Laborers from Mexico.* Hearing before the Committee on Immigration and Naturalization, 69th Congress, 1st Session. Washington DC: GPO, 1926.

———, ed. *Special Report on the Beet-Sugar Industry in the United States.* Washington DC: GPO, 1898.

United States Congress, Senate. *Smoot Hawley Tariff Debates.* 71st Congress. Congressional Record 1700. Washington DC: GPO, 1930.

United States Department of Agriculture. *Yearbook of the United States Department of Agriculture, 1901.* Washington DC: GPO, 1902.

———. *Yearbook of the United States Department of Agriculture, 1909.* Washington DC: GPO, 1910.

———. *Yearbook of the United States Department of Agriculture, 1923.* Washington DC: GPO, 1924.

United States Department of Agriculture, Office of Administrator of Research and Marketing Act, and Sugar Advisory Committee. *Sugar Problems under the Research and Marketing Act of 1946.* Washington DC: GPO, 1947.

United States Department of Agriculture, and Production and Marketing Administration. *The Pesticide Situation for 1952–53.* Washington DC: GPO, 1953.

United States Department of Commerce. *Agriculture Census of the United States: Colorado.* Washington DC: GPO, 1942, 1945, 1952, 1956, and 1961.

United States Department of the Interior, and Bureau of Land Management. *The New Empire of the Rockies: A History of Northeast Colorado.* CreateSpace, 2014.

United States Department of Labor. *Report of the Special Committee Appointed by the Secretary of Labor to Investigate Complaints against the Temporary Admission of Aliens for Agricultural Purposes.* Washington DC: GPO, 1920.

United States Tariff Commission. *Sugar.* Washington DC: GPO, 1926.

Vail, David D. *Chemical Lands: Pesticides, Aerial Spraying, and Health in North America's Grasslands Since 1945.* Tuscaloosa: University of Alabama Press, 2018.

Valdés, Dennis Nodín. "Settlers, Sojourners, and Proletarians: Social Formation in the Great Plains Sugar Beet Industry, 1890–1940." *Great Plains Quarterly* 10, no. 2 (Spring 1990): 110–23.

Valenčius, Conevery Bolton. *The Health of the Country: How American Settlers Understood Themselves and Their Land.* New York: Basic Books, 2002.

Van Den Bosch, R. *Pesticide Conspiracy.* Dorchester: Prism, 1978.

Vargas, Zaragosa. *Labor Rights Are Civil Rights: Mexican American Workers in Twentieth-Century America.* Politics and Society in Twentieth-Century America. Princeton NJ: Princeton University Press, 2005.

Walker, Richard. *The Conquest of Bread: 150 Years of Agribusiness in California.* New York: New Press, 2004.

"Water Supply of Sugar Factory." *Sugar Press,* February 1920.

Watrous, Frank. *Sugar Beets.* Bulletin no. 22. Fort Collins: Agricultural Experiment Station, Colorado Agricultural College, 1892.

Weaver, John Ernest, and Frederick William Albertson. *Grasslands of the Great Plains: Their Nature and Use.* Lincoln NE: Johnsen, 1956.

Webb, Walter Prescott. *The Great Plains.* New York: Ginn, 1931.

"Weld County, Greeley, and the Unions Colony." *Colorado Farmer,* May 8, 1979.

West, Elliott. *The Contested Plains: Indians, Goldseekers, and the Rush to Colorado.* Lawrence: University Press of Kansas, 1998.

Whitaker, James W. *Feedlot Empire: Beef Cattle Feeding in Illinois and Iowa 1840–1900.* 1st ed. Ames: Iowa State University Press, 1975.

White, Richard. *"It's Your Misfortune and None of My Own": A History of the American West.* Norman: University of Oklahoma Press, 1991.

——. *The Organic Machine.* New York: Hill and Wang, 1995.

——. *Railroaded: The Transcontinentals and the Making of Modern America.* New York: W. W. Norton, 2012.

——. "The Winning of the West: The Expansion of the Western Sioux in the Eighteenth and Nineteenth Centuries." *Journal of American History* 65, no. 2 (1978): 319–43.

Whorton, James C. *Before Silent Spring: Pesticides and Public Health in Pre-DDT America.* Princeton NJ: Princeton University Press, 1975.

Willard, James F. *Experiments in Colorado Colonization, 1869–1872: Selected Contemporary Records Relating to the German Colonization Company and the Chicago-Colorado, St. Louis-Western and Southwestern Colonies.* University of Colorado Historical Collections Vol. 3. Boulder: University of Colorado, 1926.

———. *The Union Colony at Greeley, Colorado, 1869–1871.* University of Colorado Historical Collections Vol. 1. Boulder: University of Colorado, 1918.

Williams, Hattie Plum. "A Social Study of the Russian German." Master's thesis, University of Nebraska, Lincoln, 1916.

Worster, Donald. *Dust Bowl: The Southern Plains in the 1930s.* Oxford: Oxford University Press, 1979.

———. "Good Farming and the Public Good." In *Meeting the Expectations of the Land: Essays in Sustainable Agriculture and Stewardship,* edited by Wes Jackson, Wendell Berry, and Bruce Colman, 31–41. San Francisco: North Point Press, 1984.

———. *Rivers of Empire: Water, Aridity, and the Growth of the American West.* 1st ed. New York: Pantheon Books, 1986.

———. "Transformations of the Earth: Toward an Agroecological Perspective in History." *The Journal of American History* 76, no. 4 (1990): 1087–106.

Wyckoff, William. *Creating Colorado: The Making of a Western American Landscape, 1860–1940.* New Haven CT: Yale University Press, 1999.

Wyman, Mark. *Hoboes: Bindlestiffs, Fruit Tramps, and the Harvesting of the West.* 1st ed. New York: Hill and Wang, 2010.

Zerbe, Richard. "The American Sugar Refinery Company, 1887–1914: The Story of a Monopoly." *Journal of Law and Economics* 12, no. 2 (1969): 339–75.

Zimdahl, Robert. "Pesticides: A Value Question." *Bulletin of the Entomological Society of America* (June 1972): 109–10.

INDEX

Page numbers in italics indicate illustrations.

BRDs (bovine respiratory diseases), 203, 207–8, 212, 225, 229
breeding: cattle, 41; hybrid corn, 192–94, 255–56n22; monogerm beets, 111, 157, 158; open pollination challenges for corn, 191–92
Brewbaker, H. E., 154–55, 157, 164
Brisbane, Albert, 23
Brothers Food Distribution Company, 238
Brush CO, 32–33, 61
Bureau of Agricultural Economics, 160
Bureau of Agricultural Engineering, 129. *See also* Parshall, Ralph
Bureau of Animal Industry, 160
Bureau of Entomology and Plant Quarantine (BEPQ), 177, 178, 181, 183, 190, 197
Bureau of Plant Industry (BPI), 67, 110, 111. *See also* SPI (Sugar Plant Investigations)
Bureau of Reclamation, 9, 125, 159–60, 167, 280n51
Burroughs, Wise, 209
Byers, William, 24
byproducts: beet, 67–70, *68*, 94, 183, 241; corn, *194*, 195–96, 230–32, *231*, 286n51; meat, 2, 224–25

cabbage, 43
CAC (Colorado Agricultural College): and beet growing contest, 107; collaborative relationship with USDA, 116–17; employment after graduation, 73; guidance for farmers, 186, 187, 188, 189–90; influential role in the Piedmont, 19, 42, 260n61; research on alfalfa,

44–45, 65, 261n65; research on beet byproducts, 67, 68, 69; research on beet cultivation, 57, 64–65, 80; research on biocides, 158, 183, 184, 187, 197; research on corn hybrids, 193–95; research on fertilizer, 122, 187, 265n35; research on irrigation, 129, 131, 137, 160, 164; research on machinery, 153–54; research on soil health, 64–65, 115
Cache la Poudre Irrigation Company, 32
Cache la Poudre River, 25, 26, 30, 33, 126, 130, 165
California: agricultural development in, 9–10, 254n15; beet sugar industry, 264n26; citrus industry, 262n5, 264n33; irrigation in, 33, 126; land costs, 254n13; synthetic biocides and fertilizers in, 282n7
Campbell, Thomas, 264n33
Campion, John, 47
cane sugar industry, 51, 53–56, *54*, 58, *142*, 147–48
capitalism, scholarship on industrial agriculture and, 11, 48–49, 254–55nn18–19, 261–62n4, 264n33
Capital Packing, 233
Carpenter, Louis, 116–17
Carson, Rachel, 249, 291n6
Catherine the Great, 81
Catholic charities, 104, 139–40, 145, 277n11
Catholic Conference on Industrial Problems, 149
cattle: bison compared to, 34–35; breeding, 41; diseases, 181, 203–4, 207–8, 212, 225, 229; judged competitions, 215, 219–23; in winter climate, 35, 38. *See also* beef; feed

cattle ranching and mixed husbandry: challenges on the Piedmont, 35; collapse of open-range, 37–38; Iliff's business in, 35–37; introduction to the Piedmont, 19, 34; sustainability transformations, 39–42; water supply and rights, 36–37, 39, 42, 260n59. *See also* commercial cattle feeding industry; feed

c-bt (Colorado-Big Thompson Project): benefit for farmers, 165–67, 197–98; and Colorado River Compact, 280n50; construction of, 165, *166*; and corn production, 197–99, 251; cost of, 163–64, 167, 281n58; goal of, 158–59; Great Western's promotion of, 164–65; limited federal power reflected in, 167–68; report on economics of, 160–62, *163*; violation of Reclamation's core principles, 159–60

ccfa (Colorado Cattle Feeders Association), 204, 211–12, 213, 217–18, 288n17

Cercospora (leaf spot), 116, 122, 154, 157

Chamberlain, Ray, 218

charitable organizations, 104, 139–40, 145, 277n11

chemicals. *See* biocides; fertilizers; pharmaceuticals

Cheyennes, 20, 21, 22, 57n14

child labor, *83, 97, 99;* and education, 98–100, 140; physical toll of labor, 79, 96–98; statistics, 95; and Sugar Act, 150, 151–52

Children's Bureau, 145, 149

chlordane, 180

chlorinated hydrocarbons, 176, 180, 185. *See also* ddt (dichlorodiphenyl-trichloroethane)

chlorine, 122, 124

chloropicrin, 185

citrus industry, 262n5, 264n33

climate, in Piedmont ecosystem, 21–22, 38

clover, 19, 43, 44

coal, 60, 62

Coffin v. Left Hand Ditch Co., 159

colonias (worker-owned homes), 101, 141

Colorado Agricultural College. *See* cac (Colorado Agricultural College)

Colorado and Southern Railway (c&s), 60, 61, 62

Colorado-Big Thompson Project. *See* c-bt (Colorado-Big Thompson Project)

Colorado Cattle Feeders Association (ccfa), 204, 211–12, 213, 217–18, 288n17

Colorado Fuel and Iron, 62

Colorado Gold Rush (1858), 20, 22, 34

Colorado Labor Advocate (newspaper), 144

Colorado River, 1, 158, 280n50. *See also* c-bt (Colorado-Big Thompson Project)

Colorado River Compact, 280n50

Colorado State University. *See* csu (Colorado State University)

Columbia River, 253n6

Comanches, 22

Commerce City co, 2

commercial cattle feeding industry: classification qualifications,

cultivators, mechanized, 91, 92, 111, 153, 158
Cunfer, Geoff, 255n20
Cutter Labs, 212

dams, 126, 129, 130
Davis, Frederick, 282n7
DDT (dichlorodiphenyl-trichloroethane): early applications, 176; effectiveness and persistence, 177–79; insect resistance to, 179–80, 181, 183, 190; mentioned, 279n45
DeKalb, 192–93
Deming, G. W., 184
Denver CO, 2, 20, 145
Denver Chamber of Commerce, 57, 107
Denver Pacific Railroad, 18, 24, 25
deportation, 146, 278n28
Depression, 143, 146, 151, 160–61
DES (diethylstilbestrol), 209–11, 213, 241
Deutsch, Sarah, 89, 268n7
dicamba, 247
dichlorodiphenyl-trichloroethane. See DDT (dichlorodiphenyl-trichloroethane)
dichlorophenoxyacetic acid (2,4-D), 184, 197, 279n45
Dickens, Bill, 28
diethylstilbestrol (DES), 209–11, 213, 241
Dingley Tariff (1897), 53–55
diseases: cattle, 181, 203–4, 207–8, 212, 225, 229; from drinking water, 79, 101; pesticide protection against, 176, 181, 228, 229; plant, 116, 122, 154, 157

ditch companies (mutual irrigation), 31–32, 33, 128–29, 130–31, 135, 163, 167, 258n36
Dixon, W. A., 112
Doctrine of Prior Appropriation, 9, 30, 31
Donahue, Brian, 255n20, 262n7, 266n50
Dow Corporation, 124, 175, 185, 236
Downes v. Bidwell, 56
droughts, 26, 30, 38, 158, 197
drugs. See pharmaceuticals
Duffin, Andrew, 256n24, 264n33
Duncan colin, 11, 49, 246, 255nn19–20, 261–62n4, 262n7
DuPont, 122, 123, 124, 175, 284n29
Dust Bowl, 280n47
dystocia, 212

Eaton CO, 57, 58
education, and child labor, 98–100, 140
Eli Lilly, 204, 208, 209, 210, 212, 225
Elmore, Bart, 255n18
engineers. See irrigation engineers
Ennes, Walter, 257n9
entomology, as field, 284n28. See also pesticides
EPA (Environmental Protection Agency), 224, 290n45
Estes Park CO, 172
Evans CO, 32
experiment stations. See land grant colleges
extensive cultivation, 244, 256n2

fabrication, 235
farmers. See growers

irrigation engineers: relationship dynamics with agri-state and farmers, 108–9, 137, 167–68; role of, 125–26. *See also* Parshall, Ralph
Irving, George, 190
Isaksson, Albert, 183
Isopropyl-N-Phenyl-Carbonate (IPC), 279n45

Jackson, Wes, 255n19
Jackson Ditch Company, 137
Jackson Lake, 126
Japanese Americans, 79
Jarrell, J. F., 89
Johnson-Reed Act (1924), 86

Kansas Pacific Railroad, 24, 37
Kellogg's, 233
Kemp, Frank, 154
Kingsolver, Barbara, 5
Kiowas, 22
Knights of Columbus, Mexican Welfare Committee, 104, 139–40

laborers. *See* contract laborers
land grant colleges: federal funding, 110; prestige as research institutions, 218, 289n30; uncritical promotion of chemical paradigm, 189–91, 285n40. *See also* CAC (Colorado Agricultural College); CSU (Colorado State University)
land purchases: by Great Western, 47; legislation, 18, 36; from railroads, 18, 28, 32; rising costs, 89, 126; and speculation, 18–19, 28–29; subdividing lots, 28–29, 258n29
Larimer and Weld Ditch Company, 32
laws. *See* legislation

leaf spot (*Cercospora*), 116, 122, 154, 157
legislation: on beet laborers, 150–52; on domestic and imported sugar quotas, 148, 150, 152, 161; on funding for experiment stations, 110; on immigration, 86–87; on imported sugar tariffs, 53–56, 120, 143–45, 147, 263n15; on land ownership, 18, 36. *See also* water rights
legumes, 19, 43, 44, 64, 65
Lewis, C. S., 107
lime (material), 43, 61–62
lindane, 180
Lindquist, Arthur, 181, 182
Liquid Carbonic, 236
Little Ice Age, 22, 38
living and housing conditions, 82–84, 83, 90, 100–102, 140, 150, 271n59
Logsdon, Gene, 171
Longmont CO, 32, 57, 58, 65, 67, 73, 165
Loveland CO, 47, 57, 58, 69, 107, 165
Loveland and Greeley Canal, 32
Lux, Charles, 254n15

M., Frank, 145
machines. *See* mechanization
Maddux, C. V., 88, 103, 140–41, 147
magnesium, 43
Mahony, Thomas, 139–41, 144, 146, 147, 151
manure: concentration in feedlots, 4; ddt sprayed on, 178; as fertilizer, 2, 42, 43–44, 50, 65, 66, 69; supply outpacing demand, 230; and vaccine development, 203–4; as waste, 216
Mapes, Kathleen, 263n15, 268n7

Marcus, Alan, 260n61, 289n30
Massachusetts, farming in, 262n7, 266n50
Matsushima, John, *217*; assessment of cattle feeder needs, 218–19; background, 216–17; Fed Beef Contest, 219–20, 221–23; hired at CSU, 217–18; research on cattle feeding, *222*, 223–26, 232–33
Maxfield, Richard (Dick), 183, 196
May, William John, Jr., 263n15
Maynard, Jack, 69–70; *Beets and Meat*, 70
McLeod, Asa, 258n26
Mead, Elwood, 33, 131
mechanization: of cattle feeding, 223, 226; of corn silage processing, 195–96; of harvesting, 93–94, 153, 154, 157, 270n46; incentivized by Sugar Act, 152–54; long-term impact of, 158, 169; of seed planting, 91; of seed segmentation, *155*, 156; tractors, 92, 153, 158, 183, 185, 223, 242; of weeding, 91, 92, 111, 153, 158
medicines. *See* pharmaceuticals
Meeker CO, 46
Meeker, Nathan: agricultural vision, 17, 23, 24, 27, 30, 32, 34, 42; death, 46; early challenges of Union Colony, 26; as founder of Union Colony, 4, 17–18, 24–25; newspaper work, 17, 24, 28, 46
Merck, 208, 211
mercury, 122, 124
Merleaux, April, 263n15, 268n7
Mexican Consulate, 147
Mexican nationals and Hispanos: braceros, 279n41; deportations, 146, 278n28; diet, 102, 271n59; as

essential to beet sugar industry, 79, 141; and immigration legislation, 86–87; motivations for contract labor, 89–90; as permanent underclass of laborers, 87–89; racial classification and discrimination, 86–87, 88, 98, 100, 141, 277n16; replacement of European labor pool, 85–86; terminology, 267n3; unpaid wages, 104–5
Mexican Revolution, 90
Mexican Welfare Committee of the Knights of Columbus, 104, 139–40
Michener, James, 17
Michigan, beet sugar industry, 61, 113, 124, 153, 264n26, 272n8
Mickle, Gordon, 177
migrant workers. *See* contract laborers
Miller, Henry, 254n15
mixed husbandry. *See* cattle ranching and mixed husbandry
Monfort: feeding operations, *3*, 4, 230–33; inspection and treatment of arriving cattle, 228, 229; as model of industrial agriculture, 1, 227; slaughtering and packing operations, 224–25, 233–38, *234*, *237*
Monfort, Kenny, 233, *234*, 235, 241
Monfort, Warren, 171–72, *173*, 217, 219
Montana, industrial agriculture in, 264n33
Morey, Chester, 58, 60, 61
mosquitos, 79, 94, 102
Muir, John, 47
mutual irrigation companies, 31–32, 33, 128–29, 130–31, 135, 163, 167, 258n36

mycorrhizae (fungi type), 21

Nash, Linda, 12, 268n6
The Nation (magazine), 140
National Child Labor Committee, 149
National Western Stock Show, 219,
    220–21
Native Americans. *See* Indians
NBC (National Broadcasting Com-
    pany), 161–62
NCWCD (Northern Colorado Water
    Conservation District), 167
Nebraska, beet sugar industry, 151,
    264n26
New Deal programs, 161
Newlands, Francis, 55
Newlands Act (1902), 159
New Republic (magazine), 140
New York, Fourierite colonies in, 23, 24
New York Tribune (newspaper), 17, 24
Ngai, Mae, 277n16
nitrogen: deficiency, 43, 116, 273n22;
    excess, 115, 273n22; organic resto-
    ration of, 2, 19, 43–44, 64, 71, 182;
    synthetic restoration of, 176, 185–
    86, 196–97, 265n35, 285n33
Northern Colorado Piedmont: map
    of, 8; physical geography and eco-
    system, 7, 21–22, 38
Northern Colorado Water Conserva-
    tion District (NCWCD), 167
North Platte River, 7

oats, 20, 40, 41, 43
OFM (Office of Farm Management), 115
Ohio, Fourierite colonies, 23, 24
oil and gas industry, 2, 242–44, 243
Olmstead, Alan, 264n33

open range cattle ranching, 19, 34–38.
    See also cattle ranching and mixed
    husbandry
Oregon, beet cultivation, 113
Over the Feed Bunk (magazine), 211,
    287n4
Owen, F. V., 157
Oxnard, Henry, 81

packing operations, 235–38, 237
Palm, Charles, 190
Panama Canal Zone, 129
Paris Green, 153, 265n35
Parshall, Ralph: assessment of irri-
    gation efficiency, 129–31, 162–63,
    275n49; invention of Parshall Flume,
    131–37, 133, 163, 168; support of
    C-BT, 160–62, 163, 163–64, 167–68
Pearl, Raymond, 119
peas, 43, 44, 64
penicillin, 208
Peppler, Kent, 182, 188, 198, 205, 241–
    44, 251
pesticides: alternatives to, 181–82,
    283n22; field trials, 177, 183, 225;
    insect resistance to, 179–80, 181,
    190; Paris Green, 153, 265n35;
    protection against diseases,
    176, 181, 228, 229. *See also* DDT
    (dichlorodiphenyl-trichloroethane)
petrochemicals. *See* biocides;
    fertilizers
Pfizer, 204, 212
pharmaceuticals: feedlot drug pro-
    grams, 211–12, 213; fundamental
    alteration of biological processes,
    214; funding for research, 225;
    research on and distribution of
    antibiotics, 208–9, 212; research on

seeds, corn: from hybrids, 193; from open pollination, 191
seepage, 130, 275n49
segmentation, seed, *155*, 156
Semesan, *123*, 124
Shadow Mountain Reservoir, 165
sheep, 40, *41*
Shell Corporation, 124, 175, 185, 225
shortgrass prairies, 21, 33, 34
silage: beet, 67–68, 94, *97*; corn, *194*, 195–96, 230–32, *231*, 286n51. *See also* pulp, beet
Simms, Willard, 221
Sioux, 20, 22
Skuderna, Anton, 122, 154, 155
slaughtering operations, 224–25, 233–35, *234*
Smoot, Reed, 144
Soluri, Brian, 268n6
Sonnenberg, Jerry, 291n63
Southern Colorado Piedmont, 114–18, 280n47
South Platte River, 1, 7, 8, 17. *See also* irrigation
Spain, sugar industry, 51
Spanish-American War (1898), 55
speculation, 18–19, 28–29
SPI (Sugar Plant Investigations): establishment and goals, 111; growth and influence, 114; priority shifts after World War I, 120–21; research on beet pathology in Southern Piedmont, 115–17; research on biocides, 183; research on breeding seeds, 111–13; research on fertilizer, 122–24
Spillman, William, 115–17, 273n22
Sprague, James, 218–19
Spring, Agnes Wright, 37, 259n47

Standish, Sierra, 270n40
Stanley, William, 112
state-sponsored research. *See* agri-state scientists and research
Steinberg, Ted, 254n9
Sterling CO, 32–33, 61
Stewart, Mort, 255n20, 268n6
Stilbosol, 209, *210*, 213
stirofos, 225
Stoll, Steven, 256n2, 261n4, 262n5, 264n33, 282n7
strikes contract laborers, 146
sugar: consumption rates, 51–53, *54*, *142*, 274n32; quotas on domestic and imported, 148, 150, 152, 161; tariffs on imported, 53–56, 120, 143–45, 147, 263n15; wartime supply shortage, 85, 118–20. *See also* beet sugar industry
Sugar Act (1934), 150–52, 154, 161
Sugar Beet Growers Association, 60
sugar beets: biocides, 158, 183–85, 247; blocking and thinning, 78, 91–92, 94; byproducts, 67–70, *68*, 94, 183, 241; climate and weather considerations, 51, 94–95, 103–4; and crop rotation, 63, 64–66; fertilizers, 63, 122–24, *123*, 185–86; germination, 77, *78*; harvesting, 93–94, *96*, *97*, 270n46; physical appearance, 50–51, *52*; weeding, 92–93, 94. *See also* beet sugar industry; seeds, beet
Sugar Equalization Board, 120
Sugar Plant Investigations. *See* SPI (Sugar Plant Investigations)
Sugar Trust (ASRC), 53–55, 58–60
sulfamethazine, 212
sulfuric acid, 43
Summons, Terry, 208

U.S. Land and Sugar Company, 117
U.S. Public Health and Fish and
  Wildlife Services, 176
U.S. Tariff Commission, 147–48
U.S. War Production Board, 177
Utah, beet sugar industry, 61, 122,
  124, 144, 264n26
Utes, 46

vaccines, 203–4, 228, 229
Vail, David, 189
Valdés, Dennis Nodín, 146
Valenčius, Conevery Bolt, 257n20
Van Den Bosch, Robert, 283n22,
  291n7
veterinary medicine. See
  pharmaceuticals
Vlass, Norman, 73–74

wages, of contract laborers, 82, 103,
  104–5, 121, 140–41, 143–44, 150
Walker, Richard, 261n4, 264n33
Wallace, Henry, 148, 149, 150
Washington: beet cultivation, 113;
  synthetic biocides and fertilizers in, 282n7; wheat cultivation,
  264n33
water rights: for beet factories, 60,
  164; Doctrine of Prior Appropriation, 9, 30, 31; in early settlements,
  28, 30–31, 253–54n9, 260n59; rising
  value of, 126; violation of beneficial use, 131
water sources and supply: for beet
  cultivation, 71, 126–28, 128; for
  beet sugar processing, 57, 60, 62;
  for cattle ranching, 36–37, 39, 42;
  for contract worker households,

100–101; for fracking, 242; mea-
  surements, 132–34, 253n6; on the
  Piedmont, overview, 7–9. See also
  irrigation
weeding: herbicides, 158, 184–85,
  197, 238, 242, 247–50, 291n63;
  by human labor, 92–93, 94; by
  machine, 91, 92, 111, 153, 158
weirs, 132
West, Elliott, 22, 257n9
wheat: as cash crop, 64, 254n15; early
  settler cultivation of, 20, 26, 28,
  43; as livestock feed, 40, 266n43;
  monocropping of, 264n33; surplus,
  161
White, Richard, 9
Wiley, Harvey, 80, 108
Wiley, William, 114–15
Willard, James, 32
Williams, Eugene, 257n9
Wilson, Woodrow, 120
Windsor CO, 57, 58
Wolf, Dale, 284n29
women's labor, 102
workers. See contract laborers
World War I, 85, 86, 87, 89, 118–20,
  122
World War II, 176, 208
Worster, Donald, 30–31, 139, 252,
  253–54n9, 254–55nn18–20, 258n36,
  268n6
WPA (Works Progress Administra-
  tion), 151
Wyckoff, William, 7, 253n5
Wyman, Mark, 268n7
Wyoming, beet sugar industry, 151,
  264n26

Zimdahl, Robert, 247–50, 248

Lightning Source UK Ltd.
Milton Keynes UK
UKHW010940250522
403469UK00002B/107